Earth is not a Globe in Space

The Truth About the Earth Hidden by the New World Order

By

Abraham Phoenix

Copyright © 2022 Abraham Phoenix All rights reserved. No portion of this book may be reproduced in any form without permission from the publisher, except as permitted by U.S. copyright law.

ISBN: 9798371137210

Table of Contents

Introduction 1

The Earth Is Flatter Than a Pancake 12

The Flat Earth is Proven by Airplane Level Flight 17

Horizon Is Continually Flat 28

Always Flat Horizon is At Eye Level 34

The Moon's Cold Light 50

Freemasonry and NASA 57

The Flat Earth Map 87

Crushing Gravity 99

Simple Evidence Against Heliocentrism 117

Objects in the Distance Can Be Seen Over Water 127

Introduction

This book exposes the granddaddy of conspiracies. It will present scriptural justification and indisputable proof that will make the scales fall from your eyes and make it clear that the world you had previously believed in is a fantasy. The book deals with actual facts, not made-up theories.

An honest person who is proven to be in error must either accept responsibility or stop being honest. Many so-called scientists and theologians have allowed lying to be tolerated. They have sacrificed their morality and prostrated themselves in servitude to mammon and the approval of men in order to maintain their status in life.

In a period of "universal dishonesty," according to George Orwell, "speaking the truth is a revolutionary act." Our time is one like that. The widespread misconception that the earth is a globe rotating at an average speed of 1,000 miles per hour near its equator and 66,600 miles per hour as it orbits the sun is now the most widely accepted lie. 1 The sun is thought to be hurling through the Milky Way Galaxy at about 500,000 miles per hour as the planet orbits and spins. According to some estimates, the Milky Way galaxy is moving through space at a speed of between 300,000 and 1,340,000 miles per hour.

Most people are unaware of the fact that all of this circling, whirling, and zipping across space has never been shown. These speculative motions and speeds are entirely fictitious. In truth, the results of every single scientific experiment conducted to ascertain the earth's motion have demonstrated that the planet is entirely immobile. However, the evidence that refutes the idea that the Earth revolves and orbits is ignored in scientific textbooks.

Even though the bible explicitly declares that the world does not move and is not a globe, "Christian" churches have joined the deceit and now instruct students in "Christian" institutions to view the universe as heliocentric, with the earth rotating and orbiting the sun. How was this deceit so thorough that it included both the scientific community and the "Christian" churches? Why would the scientific community support a myth that has been disproven? Why would the so-called "Christian" churches support a lie that defies the Bible? This book will provide a rationale for what occurred.

Any data that challenges the heliocentric paradigm triggers a visceral reaction in people because of conditioning. Such unjustified rejection of concrete facts results in foolishness and embarrassment. He who responds to a subject before hearing it is foolish and deserving of disgrace. (18:13 in Proverbs) This foolishness and dishonor are expressed in society's pervasive, demonic ideologies. Marxist communism, Freudian psychoanalysis, and Darwinian evolution all rest on the delusion of a spherical, rotating world. In fact, the prevalent doctrine of heliocentrism—the sun as the center of a solar system—is directly responsible for the gradual growth of the sodomite subculture into a government-protected privileged class. You query, "How so?" It is very easy.

How, one may wonder, can giving special rights to a group of sinners violate Christians' God-given, individual rights? Let's use the 2007 Oregon Equality Act as an illustration, which provides sodomites with unique state benefits. The Oregon Bureau of Labor and Industries (BOLI) fined Christians Aaron and Melissa Klein $135,000 in 2013 for breaking the Oregon Equality Act of 2007 by refusing to prepare a cake for a lesbian wedding ceremony. The penalty was assessed following a civil rights complaint made by the lesbian pair Rachel Cryer and Laurel Bowman against the bakery "Sweet Cakes by Melissa" for "emotional, mental, and bodily pain." According to Tod Starnes' account for Fox News, the couple's little bakery was destroyed by the sodomite protestors, forcing them to shut down their business:

The bakery's proprietors, Aaron and Melissa Klein, faced a lot of opposition. Gay rights organizations started picketing and protesting in front of the family business. They threatened the bakery's customers who worked as wedding vendors. Additionally, Klein informed me that the family's children had received death threats. The family's retail store finally had to close, and they now run the bakery out of their house. They issued a message pledging to maintain their religious convictions.

"I honestly didn't intend to harm anyone or offend anyone," Melissa Klein said. "It's simply something I believe in very passionately." Aaron Klein had to work as a trash collector to make ends meet when the Kleins were forced to close their bakery. The law of Oregon's state "frameth mischief" against the soul of the just. It is not unusual, as in the Klein case in Oregon.

All around the nation, Christian rights are being curtailed. Elaine and Jonathan Huguenin, two Christian photographers, were found to have violated the state's human rights law by refusing to take pictures of a same-sex union, according to a majority decision by the New Mexico Supreme Court. The Washington State Attorney General brought legal action against a florist who turned down a same-sex couple's request for flowers. A judge ordered Colorado's Masterpiece Cakeshop owner Jack Phillips to serve sodomite couples or face penalties. "When the upright have power, the people celebrate; but when the wicked hold power, the people lament." (Proverbs 29:2)

Sin is promoted and protected, which degenerates society. As the perversions descend, one vice leads to another. The American government announced a new policy on July 19, 2016, enabling transgender people to join the military. According to a new sexual orientation military handbook, no otherwise eligible service member may be involuntarily separated from duty, dismissed, or refused re-enlistment or continued service based simply on their gender identification. The guidebook continues by outlining the steps required for military personnel to get sex exchanges paid for by the government.

Lesbian, homosexual, bisexual, and transgender (often known as LGBT) communities have a more sinister goal that they keep hidden from the general population. The dark truth of the LGBT rights movement is revealed in Michael Swift's 1987 Gay Manifesto. The LGBT community wants to have access to kid sodomization. Brazenly, the Gay Manifesto declares:
We will sodomize your boys, who serve as symbols of your weak manhood, limited aspirations, and disgusting falsehoods. Everywhere there are guys with other men, including your schools, dorms, gymnasiums, locker rooms, sports stadiums, seminaries, youth organizations, movie theater restrooms, army barracks, truck stops, all-male clubs, and Congressional chambers, we will entice them. Your boys will serve as our servants and carry out our orders. We'll remake them in our likeness. They'll start to lust after and admire us.

Steve Baldwin, a former member of the California State Assembly, conducted research on the link between homosexuality and pederasty.
His research was published in the Regent University Law Review in the spring of 2002. Scientific research, according to Baldwin, "confirms a high pedophilic propensity among homosexuals." One research from

1988 found that 86% of pedophiles who preyed on boys "described themselves as gay or bisexual," according to an article published in the Archives of Sexual Behavior.
Further research has revealed that the rate of child molestation by gays is ten to twenty times higher than that of heterosexuals. The homosexual community is aware of those realities. The Sentinel, the major gay newspaper in San Francisco, declares unequivocally that "[t]he love between men and boys is at the root of homosexuality.

Publications for homosexuals openly encourage pederasty and frequently feature travel advertising for sex trips to Thailand, Burma, the Philippines, Sri Lanka, and other nations notorious for boy prostitution. The most well-known travel book for gay people, Spartacus Gay Guides, is full of information about where to locate boys for sex and, as a kindly warning, gives punishments in several countries for sodomy with boys if found, according to Baldwin.

According to Baldwin, "sex with minors is frequently encouraged in the mainstream gay society. Leaders in the homosexual community frequently defend the right to have consensual relations with youngsters. He concluded that the legalizing and promotion of child molestation is one of the main goals of the LGBT rights movement. In order to legalize pedophilia, mainstream LGBT groups like the National Coalition of Gay Organizations and the International Lesbian and Gay Association (ILGA) have approved many organizational resolutions asking for reducing or abolishing the legal age of consent for sexual activity. More than 620 member groups make up the ILGA. Due to political pressure from the U.S. Congress and the discovery that many of the ILGA's member groups supported pedophilia, the UN decertified the ILGA's consultative status as a Non-Governmental Organization (NGO) in 1995. ILGA cut links with the North American Man Boy Love Association (NAMBLA), whose members engage in pederasty, in an effort to get restored. Breaking connections with NAMBLA was obviously only window dressing, as the ILGA continued to support a dozen more potent (and covert) pro-pedophile organizations. The UN has continued to forbid ILGA from having consultative status.

It is nothing new for sodomites to practice widespread pederasty. And the sodomite perversion is a result of the demonic spiritual influence. For instance, Aleister Crowley, a high Freemason and Satanist, openly claimed to be "the wickedest man on earth" and went by the moniker "The Beast." Crowley used tantric sex magic, often known as magick. The central strategy of tantric sex magic is to do progressively heinous and repulsive sexual activities, which eventually lead to gay behavior and the rape of a kid. Only the most elite members of Freemasonry are privy to the terrible masonic secret that the only route to immortality is to sexually abuse a kid. That is the rationale behind the government protecting sodomites' rights. The degenerative objective of the demonic occultists who rule the governments of the globe is to legalize sex with minors, and that government-protected privilege is merely the first step in that direction. The author of Antichrist: The Beast Revealed describes the various forms of pederasty that the wealthy elite of today's society indulge in in secret.

Sin feeds on sin. Once a sin is sanctioned and protected by the power of the state, it moves down a downward spiral where other sins are sanctioned and protected. The result is a society that is aggressive and disorderly. New restrictive laws are created as a result of this disorder, which ultimately results in the establishment of a police state to restore order. The exercise of God-given rights is prohibited by the state under the increased police powers provided to it.

That is not exaggeration. The U.S. Department of Justice intends to extend the "civil rights" accorded by the U.S. Government to include a slew of new perversions, including but not limited to bestiality, polygamy, and pedophilia, according to an interview with former U.S. House of Representatives Majority Leader Tom DeLay on The Steve Malzberg Show.

Christians who use their God-given rights, denounce these offenses, and work to safeguard the children will be persecuted. However, those abhorrent "civil rights" to rape children, engage in polygamous marriages, and have sex with animals will also be used as a weapon. In order to utilize the power of the state to punish rebellious Christians, the U.S. Justice Department has been busy formulating strategies.

Former U.S. House Majority Leader Tom DeLay asserts that the Justice Department has written a report outlining a dozen "perversions" that it wants to have legalized, including bestiality and pedophilia.

"We've discovered a covert document issued by the Justice Department. They will now pursue 12 other perversions. On "The Steve Malzberg Show" on Tuesday, DeLay discussed topics such as bestiality, polygamy, and having sex with young boys while making it legal "newsmax TV

They also have a long list of tactics to use against churches, pastors, and any companies who want to defend their right to practice their religion. This is approaching, and it is approaching like a tsunami.

Four days after the Supreme Court issued a momentous judgment that same-sex marriage is now legal in all 50 states, something DeLay vehemently disagrees with, the Texas Republican dropped a stunning revelation.

DeLay reacted when Steve Malzberg reiterated his claims that the Justice Department wants to "legitimatize or legalize" behaviors like bestiality, which is defined as sex between humans and animals: "That's accurate, that's correct. They have 12 fresh perversions on the way. Lesbian, gay, bisexual, and transgender is only the start of the LGBT movement. They'll start adding new perversions after they've done so.
(Parenthesis in original text are in brackets)

Satan is aware that if man indulges his appetites without respect for God's precepts, the resultant immoral behavior will call for a despotic government. How so? In essence, the increasing sin is an intrusion on someone else's rights, which are conferred by God. Jesus clarified the idea to the man, telling him that he was to love the Lord his God with all of his heart, soul, and mind. The first and most important commandment is this. The second is similar to it: You must love your neighbor as yourself. All of the law and prophets hang on these two precepts. Matt. 22:37–40

The New World Order is what Satan's servants refer it as, and it is his plan to rule over all of humanity. The former president George H.W. Bush is one of Satan's henchmen. President Bush openly proclaimed his plans to establish a New World Order in his State of the Union address on January 29, 1991, before a joint session of Congress. "A New World Order, where several nations are brought together in single cause, is at risk, not just one little country.

According to Isa Blagden, "if a falsehood is only written enough times, it becomes a quasi-truth, and if such a truth is repeated enough times, it becomes an article of belief, a dogma, and men will die for it. Today's world is ruled by the concept of heliocentrism. The heliocentric approach serves to conceal God's presence. This book will not only disprove heliocentrism, but it will also demonstrate that, as the Bible claims, God created the earth as the fixed, flat center of his whole cosmos. You'll realize that God exists and is keeping an eye on you.

1 The Earth Is Flatter Than a Pancake

Of fact, neither land nor sea are exactly flat due to their numerous mountains and valleys. However, it can be shown that land is not convex, as would be expected on a spherical earth, when huge expansive regions of land are taken into account. The land is often level. Scientific study backs up that assertion. For instance, Mark Fonstad, William Pugatch, and Brandon Vogt, Ph.D., used data from the United States Geological Survey to establish that Kansas is essentially flatter than a pancake when viewed from a distance. The geographers determined that a pancake had a measured flatness of .957 on a scale where one (1) is absolutely flat using a confocal laser. A 1:250,000 scale digital elevation model was used to scale down the State of Kansas (DEM).

The measured flatness of Kansas was reported to be .9997.

Fonstad et al. contrasted the east-west profile of combined relief data from the State of Kansas with pancake transsections.

The pancake that Fonstad and colleagues utilized came from an IHOP restaurant.

Over a 130 millimeter diameter, its relief was found to be 2 millimeters.

A landscape's vertical elevation shift over a certain region is measured quantitatively as relief. The relief of a piece of land is calculated by deducting its greatest height from its lowest elevation. Divide the relief by the length of the transection to compare the relief of two transected profiles of varying sizes. The relief of the two transected profiles may be compared using the resultant relief quotient. The region is flatter the lower the relief quotient. In the study endeavor by Fonstad et al., the relief quotient for the pancake was .015 (2 130=.015).

The state of Kansas has a maximum elevation of 4,039 feet above sea level and a minimum elevation of 679 feet. Therefore, Kansas has a 3,360 foot relief (.64 miles). Kansas spans 400 miles spanning from east to west, giving it an estimated relief quotient of .0016 (.64 x 400 =.0016). The results reported by Fonstad, et al. are supported by a comparison of relief quotients. By far, Kansas is flatter than a pancake.

Joshua Campbell, a geographer and GIS architect in the Office of the Geographer and Global Issues at the U.S. Department of State, and Jerome Dobson, president of the American Geographical Society and professor of geography at the University of Kansas, came to the state of Kansas's defense. They didn't want visitors to believe Kansas was uninteresting and flat.

Dobson and Campbell came to the conclusion that, according to the research study of Dr. Fonstad, et al., Kansas would need to have a mountain that is 32,506 feet (roughly 6 miles) above sea level (400 miles .015 relief quotient for a pancake = 6 miles) in order for Kansas to not be flatter than a pancake over its 400-mile span. 80 Such a six (6) mile high mountain would be taller than Mount Everest, which is the highest peak in the world at 29,029 feet above sea level, and around ten times the actual variance in topography in Kansas.

Kansas would be elevated more than 26,666 feet (5 miles) above sea level if the planet were a globe. It's interesting to note that Kansas would have a relief quotient of .0125, which would still make it flatter than a pancake, which has a relief quotient of .015. That implies that the earth is a sphere, right? Not at all, no.
Because Kansas only has a maximum real relief of 3,360 feet and an actual relief quotient of .0016, there isn't the five-mile bulging arc that would be necessary on a globe.

Additionally, the predicted arc of the hypothetically spherical globe would be bigger the larger the land mass. For instance, Florida could only be flatter than Kansas if the world was flat, as found by Dobson and Campbell to be the case. The length of Florida, from north to south, is 447 miles. Florida would have a bulging arc of 33,153 feet (6.27 miles) and a relief quotient of .014 if the earth were a spherical. Accordingly, Kansas, which has a relief factor of .0125 on a hypothetical spherical earth, ought to be flatter than Florida. Florida is really flatter than Kansas. Florida is, in fact, the most level state in the union.

Since Florida was shown to be flatter than Kansas using accurate measurements, the planet must be flat rather than a globe.

The real relief quotient tends to be flatter the bigger the area measured on earth, which is the exact reverse of what would be anticipated on a spherical globe. In fact, there is a significant disparity between the relief quotients predicted by a globular hypothesis and the actual relief quotients when relief quotient computations are done over a vast, thousands-mile-long area of land.

For instance, the length of the continental United States is around 2,800 miles. The continental United States would have a bulged arc across it that is 1,306,666 feet (247 miles) above sea level if the planet were a globe. There is no such topographical bulge. The continental United States should have a relief quotient of .088 (247 x 2,800 =.088) if the earth were a globe. However, the relief quotient of the United States' continental territory is well below the relief quotient (.088) that would be predicted on a spherical earth.

The continental United States' highest point rises 14,494 feet above sea level, while its lowest point is 282 feet below it. The relief is consequently 14,776 feet (2.8 miles) (14,494 + 282 = 14,776) (14,776 feet = 2.8 miles) across the 2,800 mile span of the continental United States. The 282 foot elevation is included to the 14,494 foot total because it is below sea level. A relief quotient of .001 is obtained by dividing 2.8 miles by the 2,800 miles that make up the continental United States (2.8 2,800 =.001).

The earth cannot be a sphere since the real relief quotient for the continental United States is .001. The relief quotient for the continental United States would be significantly higher if the earth were a spherical (.088).

otherwise without Compensation, in one hour at the time the pilot would do They find themselves at an altitude of 166,666 feet (31.5 miles). than expected! An airplane traveling at a typical speed of 35,000 Feet like to maintain this height at The upper edge of the so-called "troposphere" in one An hour they will find themselves over 200,000 feet Rising in the Mesosphere steadily Raise the track the longer the duration. Several pilots I spoke with confirmed that no such adjustment is ever performed for the Earth's alleged curvature. When a pilot sets an altitude, both their fake horizon gauge and course stay level.

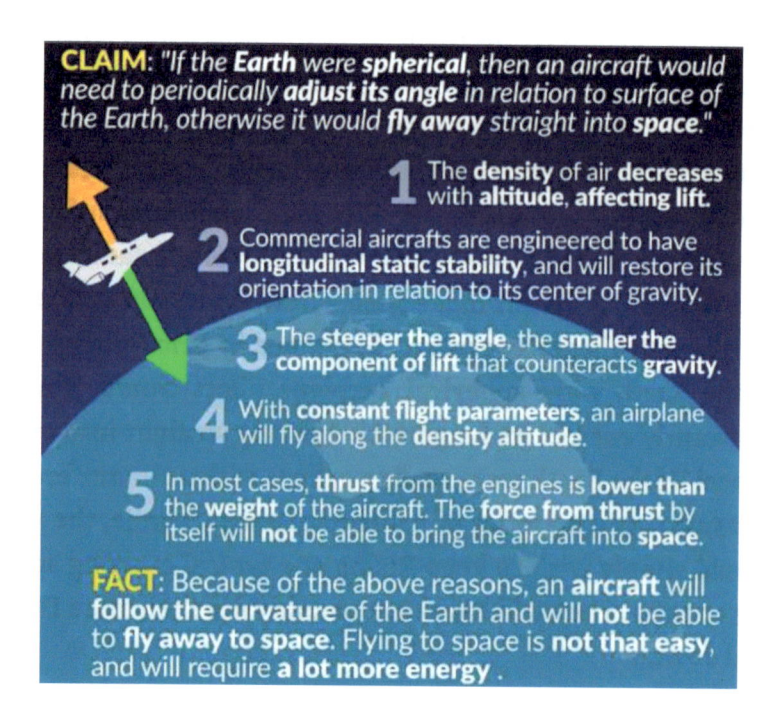

CLAIM: "If the **Earth** were **spherical**, then an aircraft would need to periodically **adjust its angle** in relation to surface of the Earth, otherwise it would **fly away** straight into **space**."

1 The **density** of air **decreases** with **altitude, affecting lift.**

2 Commercial aircrafts are engineered to have **longitudinal static stability**, and will restore its orientation in relation to its center of gravity.

3 The **steeper the angle**, the **smaller the component of lift** that counteracts **gravity**.

4 With **constant flight parameters**, an airplane will fly along the **density altitude**.

5 In most cases, **thrust** from the engines is **lower than** the **weight** of the aircraft. The **force from thrust** by itself will **not** be able to bring the aircraft into **space**.

FACT: Because of the above reasons, an **aircraft** will follow the **curvature** of the Earth and will **not** be able to **fly away to space**. Flying to space is **not that easy**, and will require **a lot more energy** .

In fact, the Federal Aviation Administration (FAA) uses its Target Generation Facility to teach pilots and air traffic controllers, assuming that the globe is flat (TGF). The TGF is made up of a number of software packages that manage simulation situations using fictitious airplanes. Nearly all of the air traffic control laboratories at the FAA William J. Hughes Technical Center are operated by the TGF computer simulator. According to the FAA, "Our lab has closely collaborated with the TGF group to make airplanes behave as air traffic controllers would anticipate them to in the actual National Airspace System" (NAS). The virtual airplanes in TGF are very accurate replicas of their physical counterparts.

To effectively instruct pilots and air traffic controllers on aircraft behavior in flight, the TGF's software must be correct. The Engineering Analysis and Design of the Aircraft Dynamics Model For the FAA Target Generation Facility [TGF], a paper from the FAA, makes the premise that the world is flat. The program that the TGF use to precisely model the behavior of airplanes over the earth is described in that FAA paper.

This work focuses on an Aircraft Dynamics Model (ADM) suited for integration into the FAA TGF simulations at the FAA William J. Hughes Technical Center, Atlantic City, NJ. It describes the comprehensive technical design and software implementation of the ADM.

Higher fidelity will be needed when new simulations are created or presented to the Technical Center to uncover NAS operational safety and performance concerns.

To fulfill the demands of the other FAA projects and simulators, the TGF is poised to improve its quality and operational connection. This project's objective has been to create and maintain a high-fidelity simulation capability that will suit the FAA's operational, testing, and evaluation demands for its NAS. (I added emphasis.)

The world is motionless and flat, and the TGF program is loyal to this fact. It would be dangerous for airplanes if the TGF simulator built its model on the erroneous assumption of a spherical, spinning earth. Since this is fact, the FAA was forced to make this assumption. The FAA would have taught pilots and air traffic controllers incorrectly and produced the messy and unpleasant consequence of frequent jet crashes if it had assumed a rotating, spherical earth model.

The algorithms in the TGA simulator that presume a flat, motionless earth have to be verified with real-world testing by the FAA since safety must always come first. "The paper finishes with a section on verification and validation, the procedure by which the various components of the simulation are evaluated and validated," according to the FAA article. What was revealed in the verification section? The testing carried out to verify and validate the TGF simulation "gives us a high degree of confidence that the models provided herein have adequate fidelity for use as a target generating tool," according to the FAA article.

The models were predicated on a flat, motionless world, which the tests strongly suggested were accurate to reality. This indicates that the FAA has a high degree of confidence that the planet is flat and stationary because of real testing with airplanes.

In another official document, the U.S. government impliedly recognized the existence of a flat, stationary earth. Derivation and Definition of a Linear Aircraft Model, a NASA article from 1988, notes that the widely used linear airplane model is based on "a rigid aircraft of constant mass flying over a flat, nonrotating earth."

Every military and commercial aircraft is equipped with an instrument that can only function on a flat earth. The attitude indicator, commonly referred to as an artificial horizon, is that device.

Even if the horizon is blocked by bad weather or nighttime, the artificial horizon instrument features a display that shows the pilot the plane's attitude to the actual horizon. The pilot can tell if the plane is rolling (i.e., flying level) and pitching (i.e., whether the nose of the plane is pointing above or below the horizon). The gyroscope placed on a gimbal and rotating is how that attitude indicator functions.

A gyroscope has two crucial characteristics: precession and spatial stiffness. These two characteristics are the foundation for all of the practical uses of gyroscopes. The gyro will resist any force that tries to alter its rotational plane due to precession.

The gyro will move in response to an applied force, but not in the direction of the applied force. Instead, the gyro will rotate at a right angle to the force's direction. The amount of force being applied directly affects the rate of precession. Rigidity in space is a gyroscope's most crucial feature for a pilot. This means that a gyroscope will maintain its vertical attitude in regard to its rotor's axis and its horizontal attitude parallel to the direction of the rotor's spin. The vertical y axis serves as the rotor's axis of rotation, while its horizontal plane is located on the x (lateral) and z (depth) axes. The gyroscope will maintain its initial attitude with respect to the x, y, and z axes if it is mounted on a gimbal that enables rotation along all three of these axes (x, y, and z), while the gimbal will rotate around the stationary gyroscope. The gyroscope's x, y, and z axes will continue to be rigid in space, providing a fixed matrix on which the airplane may travel. The attitude indicator instrument's coordinates will reflect any discrepancy between the fixed axes of the gyroscope and the moving airplane. The following frames from a 1960s U.S. Navy training video describing the use of gyroscopic flight instruments show how this is made possible by the gimbal mounted to the aircraft in the instrument bay.

The attitude indicator's rotating gyroscope remains stable and stiff in space. The aircraft can move with respect to the fixed axes (x, y, and z) of the rotating gyroscope thanks to the gimbal in the attitude indication instrument. As a result, the pilot may determine the precise attitude of his aircraft with respect to the horizon by gazing at the attitude indicator device.

The drawback of the spherical earth model is that an airplane with an attitude indicator indicating level and straight flight will lead the aircraft to fly off into the upper atmosphere as the curvature of the globe recedes from the aircraft. On a flat earth, a pilot of an airplane can only use his attitude indicator to fly a straight and level course.

An image from a 1960s United States Navy training video is shown below. 130 The frame serves as an illustration of how impossible it is for an airplane employing gyroscopic equipment to maintain level flying on a spherical earth. As it theoretically navigates the globe, the gyroscope pictured in the frame is observed to retain its rigidity in space. A plane equipped with a gyroscopic attitude indicator would progressively be moving out toward the upper atmosphere as it traveled around the world, as is evident from the gyroscope's stiff attitude.

In fact, as shown in the image below, the plane would be on a vertical trajectory after flying around 6,250 miles, or one-quarter of the earth's diameter.

By claiming that the attitude of the gyroscope integrated into the attitude indicator instrument is constantly being adjusted, the training video narrator attempted to explain away this apparent flaw. However, the fabled on-the-spot gyroscope adjustment is not possible since it would render the gyroscope worthless. Remember that a gyroscope's ability to retain its rigidity in space is its key characteristic. If a gyroscope's basic characteristic, rigidity, can be changed so simply, rigidity cannot be considered a core characteristic of a gyroscope.

If it were feasible to change the gyroscope's fundamental property of spatial rigidity, it would negate the entire purpose of using the gyroscope. It would be absurd to employ a gimbaled gyroscope in a device that calls for a secondary function to counteract the primary gyroscope feature (rigidity). The instrument's purported neutralizing function would have to determine the exact curvature of the allegedly ball-shaped earth and keep a correct course over it. Why would it need to have a gimbaled gyroscope if it could already perform all of that? On a ball earth, an attitude indicator with a gimbaled gyroscope would only be a problem.

In contrast to what is required in an attitude indicator on a spherical earth, rigidity in space is the exact opposite. On a flat earth, a gyroscopic attitude indication device is required, but on a spherical earth, the same equipment would be useless and potentially harmful.

Because the earth is flat, a gyroscopic attitude indicator maintains its attitude precisely perpendicular to its axis, on a fixed plane to the spin of its rotor, and absolutely level in regard to the earth's surface. A moving gyroscope that maintains its attitude is rigid in space, but its rigidity is only helpful as an attitude indicator on a flat earth. On a globe, a gyroscopic attitude indicator will simply not function.

The United States Navy Training Film's above graphic reveals an intriguing fact: if the planet were a spinning globe, a gyroscope in a gimbal set on a level table should display the earth's spin.

As a result, the gimbal that surrounds the gyroscope would display movement of 1.25 degrees every five minutes and 15 degrees every hour if the world were to be rotating. According to the United States Navy Training Film, the gimbal should rotate 90 degrees beneath the gyroscope on a rotating globe after six hours. A gimbaled gyroscope has been used in such an experiment. The experiment demonstrated that the earth is not a rotating globe by showing absolutely no movement of the gimbal around the gyroscope.

3 Horizon Is Continually Flat

Once someone understands what they are seeing around them, the evidence of a flat earth is simple to identify. Over water, the horizon is always perceived as being flat. That demonstrates that the world is flat since the horizon would be bent otherwise. According to B. Charles Brough:

Since the sea horizon always seems to be and actually is a perfectly level line, regardless of the angle from which it is viewed, its surface must also be level, and since this appearance is consistent across the planet, the Earth is a plane. This may be demonstrated by setting up a properly leveled board or a string that is firmly stretched between two vertical poles and is at right angles to a plumb line at an appropriate elevation on the seashore. Throughout a distance of 20 miles, the horizontal line may be seen clearly when facing the sea, and for its whole length, it coincides with the straightedge, or string: But if the world were a globe, the horizontal line would curve both directions from the center at a rate of eight inches multiplied by the square of the distance, creating an arc twenty miles long. As a result, the horizontal line in the middle of the distance should be 66 feet below the horizon at each end. However, since such an appearance is never offered, it follows that the earth cannot be a globe or anything else than a plane.

Samuel Rowbotham demonstrated that Brough's assertion was accurate. To quote Rowbotham:

It is well known that the horizon at sea always looks as a straight line, regardless of how far to the right or left of the observer on land it may reach. The following experiment has been conducted across the nation. Two poles were buried in the ground in Brighton, six yards apart and immediately across from the sea, on an inclining area close to the racetrack. A line that was rigidly stretched parallel to the horizon ran between these poles.

At least 20 miles were visible from the line's center on each side, for a total of 40 miles. A ship was seen going due westward; for several hours or until the ship had traveled the entire 40 miles, the line cutting the rigging just above the bulwarks was visible. The ship entering view from the east would have to travel 20 miles up an incline before reaching the arc's center, where it would then have to go back down the same distance. The amount the vessel would be below the line at the start and end of the 40 miles is calculated by multiplying the square of 20 miles by 8 inches, which results in a value of 266 feet.

In fact, Rowbotham provides vivid evidence that the planet is flat, which may be seen by just glancing at the sea horizon. Anyone may duplicate his proof.

The sea horizon always looks as a perfectly straight line, as shown by H, H [in the image below], no matter how far to the right and left of an observer on land it stretches.

Not only does it seem to be straight as far as it goes, but the next easy experiment may demonstrate that it is. Fix a long board edgewise on tripods at any elevation above sea level, as illustrated in [the image below]. The board should be between 6 and 12 feet long or longer.

Make sure the upper edge is level and smooth. The distant horizon can be seen to run precisely parallel with the upper edge of the board B, B, when the eye is placed behind and roughly in the center of the board and directed over it towards the water. According to the altitude of the place, it will not be difficult to observe a length of ten to twenty miles if the eye is now pointed in an angular orientation to the left and to the right; the entire distance of twenty miles of the sea horizon will be perceived as a completely straight line. If the planet were a globe and the ocean's water convex, this would not be conceivable. Ten miles on either side would result in a 66-foot curvature (102 times 8 = 66 feet 8 inches), and rather than the horizon touching the board along its entire length, it would be seen to gradually decline from the center C and to be more than 66 feet below the two extremities B, B, as shown in fig. 18. Any ship coming from the left would be observed to climb the inclined plane H, B, C, and after passing through the middle, it would fall from C towards the sloping horizon at H. Since this phenomena has never been seen, it is reasonable to assume that there is no such convexity or curvature.

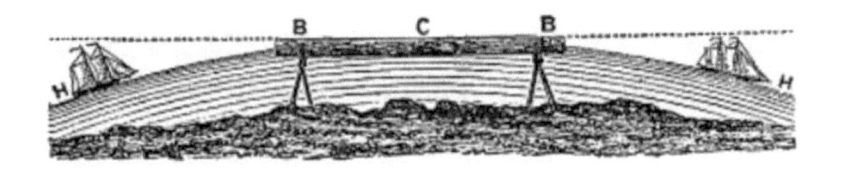

Rowbotham uses observation of a certain British shoreline to further support his claim that the globe is flat. His observations on the North Wales coastline's flatness may still be verified today. Today, anybody may verify that the planet is flat by looking at the coastline of North Wales from the Isle of Man, much as Rowbotham did in 1881.

The whole length of North America's coast, starting from the high ground at Douglas Harbour on the Isle of Man Wales may frequently be seen with the unaided eye from Ayr, at the mouth of the River Dee, towards Holyhead, a distance of at least fifty miles. No matter the method used, it has always been discovered that the line connecting the sea and the land seems to be precisely horizontal, as illustrated in the accompanying picture (Fig. 21). However, if the earth were spherical and all water's surface was convex, such an appearance could not exist. It would unavoidably look as fig. 22 depicts. Along with displaying the different land elevations, a horizontal line stretched in front of the observer would also display the declination of the horizon H, H, which is below the cross-line S, S. The Welsh coast, visible along the horizon in Liverpool Bay for fifty miles, would be at least 416 feet away from the center (252 x.8 inches = 416 feet 8 inches). However, because such a declination or downward curvature cannot be seen, The conclusion that it doesn't exist is logically necessary.

Let the reader consider if there is any natural explanation for why a fall of more than 400 feet is not visible to the naked eye or detectable by any optical or mathematical techniques. This topic is particularly significant in light of the fact that variations in level of only a few yards extent are immediately and undeniably evident at the same distance and on the upper outline of the same area. He will no longer be able to claim that the world is a globe if he is guided by logic, reason, a love of truth, and consistency.

He must believe that to do so is to reject the evidence of his senses, dismiss the significance of fact and experiment, completely discount the usefulness of reasoning, and stop relying on practical induction.

Always Flat Horizon is At Eye Level

No matter how high the observer ascends, the horizon is never just flat to the observer but also always climbs to eye level with the observer who takes flight. If the world were a globe, the observer's eye level would be lowered as he rose above the planet, and the lateral horizon would be bent. Explained by Eric Dubay:

The eternally horizontal horizon line always comes up to meet the spectator's eye level and remains perfectly flat, whether the observer is on the surface of the ocean, at the summit of Mount Everest, or flying at altitudes of over 100,000 feet. Test it out for yourself on a beach or a mountaintop, in a big field or the desert, in a hot air balloon or a chopper, and you'll notice that the panoramic horizon rises with you and stays perfectly level all the way around. The horizon, however, should descend as you ascend rather than rise to your eye level. It should also descend at either end of your peripheral rather than remain level all the way around if the Earth were genuinely a large globe. The highest point of the ball-Earth would be right under you and dropping on each side if you were standing in a rising balloon and looking toward the horizon. When J. Glaisher looked over the top of the car and took a broad view of the entire visible area below, he observed that "the horizon appeared to be on a level with the eye, and taking a grand view of the entire visible area beneath, I was struck with its great regularity; all was dwarfed to one plane; it seemed too flat."

According to Samuel Rowbotham, if the earth's surface were convex, as it would need to be on a globe, a balloonist would see the horizon slip down as the balloon ascended rather than remaining at eye level. This would force the balloonist to look down in order to see the horizon. 137 However, the balloonist does not observe that. The horizon is always visible to a balloonist at eye level.

The horizon would have been 127 miles away, more than 10,000 feet below the summit of the arc of water beneath the balloon, and more than 20,000 feet below the line of sight A, B, as shown in [the below figure]; and the dip C, H, from C, B, to the horizon H, would be so great that the astronaut could not fail to observe it; instead of which he always sees it "o" Rowbotham demonstrates this point in the image below.

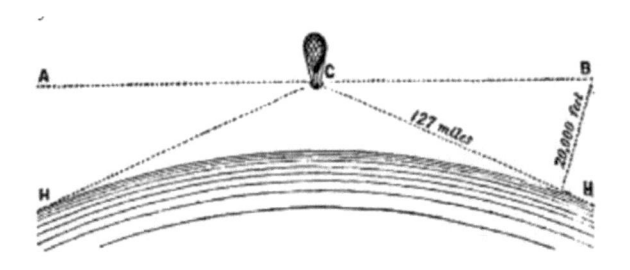

Any passenger on a commercial aircraft nowadays has the opportunity to witness the exact same occurrence as a balloonist. From takeoff to the plane's cruising height of 30,000 feet, the earth's horizon is always visible at eye level. The horizon would be below the observer's point of view if the earth were a globe. The world is a flat plane because the horizon is always visible to the passenger at eye level. Eric Dubay provides an explanation of the recent emergence of photographic trickery employed to produce curved earth representations in **NASA** and other high altitude images.

You may view videos online showing the horizon rising with the camera level and remaining flawlessly flat 360 degrees around. Amateurs have launched balloons to altitudes of over 121,000 feet.
However, wide-angle lenses and post-production work have been used in **NASA** footage and other "official" sources to impart artificial curvature to the Earth, including the recent Red Bull skydiving at 128,000 feet. Panoramic images taken from the summit of Mount Everest also frequently purport to show the curvature of the Earth, although these images are just distorted due to the limits of wide-angle lenses.

The two images of the world below provide evidence for Dubay's claim that **NASA** and the controlled media misrepresented a spherical earth. NASA Commander Scott Kelly allegedly took the photo at the top when he was on board the International Space Station (ISS). 140

The image below was most likely obtained from a high altitude airplane and does not, virtually surely, depict the space-based ISS. The top image features an inset of the camera and wide-angle, fish-eye lens that Commander Kelly employed; Commander Kelly provided the image of the camera. The software will then automatically adjust the distortion brought on by the fish eye lens. Adobe Photoshop has a feature that allows one to input the type of camera and type of lens. A photographer who is skilled in this technique achieved the absolutely level horizon shown in the bottom image below.

Take note of the sun's hot spot, which shows that the sun is directly above and considerably smaller than the earth.

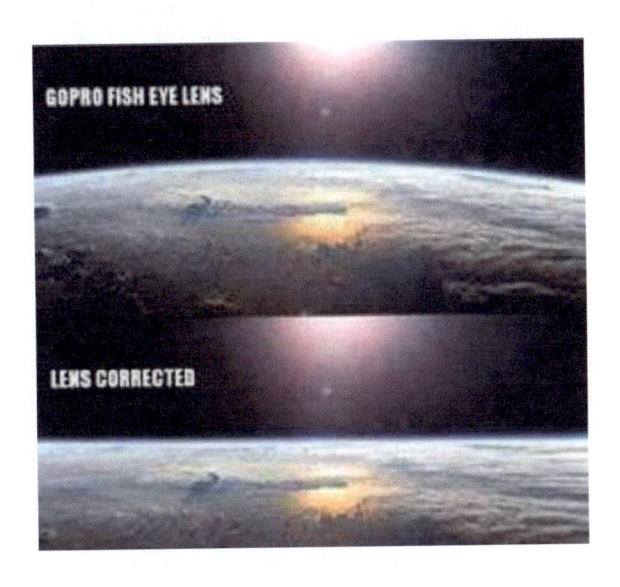

By viewing Felix Baumgartner's Red Bull-sponsored world record skydiving, Dubay's claim that NASA and the controlled media's representations of a spherical earth are false may be further corroborated. A camera with a regular lens captures the flat, eye level horizon of the earth before Baumgartner stepped out of his capsule supported by a balloon at around 128,000 feet above the surface. Because the photo was shot just before the leap, it shows a height of 127,518 feet; nevertheless, the balloon continued to rise until the jump height of more than 128,000 feet. At 127,518 feet above the surface of the earth, the horizon shouldn't have been visible to the naked eye if the planet were a globe.

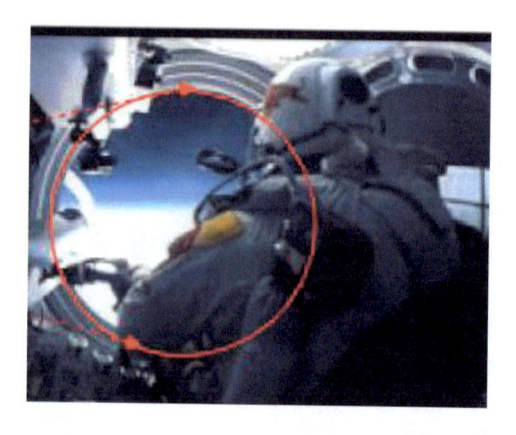

The level horizon abruptly changes to a curved horizon when Baumgartner exits the balloon's capsule and the film is switched to the external GoPro camera. The outside GoPro camera employed a wide-angle, fish-eye lens, which made the horizon of the world seem bent. This is how the curved horizon is seen.

The image below was taken not long after Baumgartner leaped off the spacecraft. Take note that the GoPro camera's fish-eye lens depicts a concave earth. Depending on how the camera is pointed towards the horizon, the fish eye lens will depict the earth's horizon as either concave or convex. The level horizon in the image below seems concave because the earth is near the outside edge of the fish-eye lens. However, in the image above, the sky is near the edge of the fish-eye lens, making the flat horizon of the earth look convex. The horizon of the earth is actually flat, neither convex or concave.

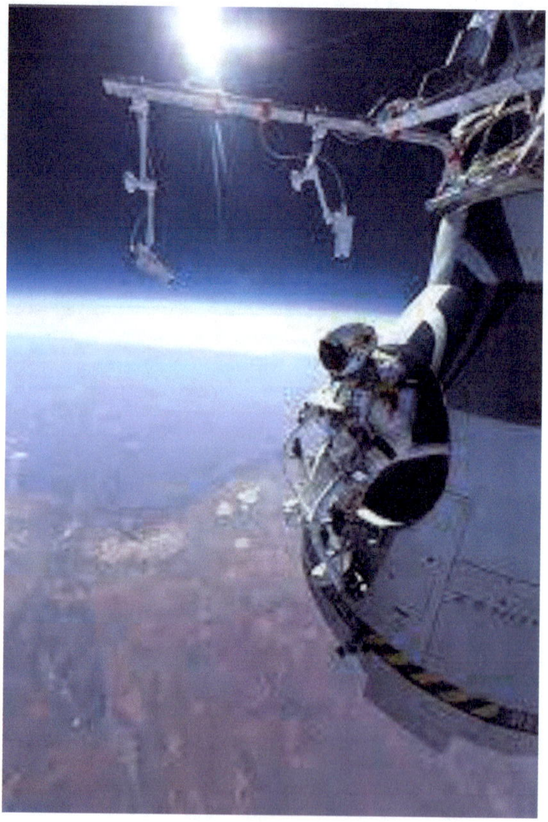

> Earth Level
>
> Sky level
>
> Sea level
>
> Horizon always at eye level is only possible on a flat plane.

A stationary earth cannot be shown by the Coriolis effect.

One rule of motion on a rotating globe is that the Coriolis effect, as it is known, will unavoidably be produced by the spinning. Gustave-Gaspard Coriolis, a French engineer, mathematician, and physicist who was born on May 21st, 1792, and passed away on September 19th, 1843, was the one who originally proposed the Coriolis effect. The Coriolis force is described as "an effect of motion on a spinning body, of essential importance to meteorology, ballistics, and oceanography" by the Encyclopedia Britannica.

The Coriolis force is further explained by The Encyclopedia Britannica in relation to the allegedly rotating earth as follows:
In 1835, Coriolis published a paper titled "Sur les équations du mouvement relatif des systèmes de corps," in which he demonstrated that on a rotating surface, in addition to a body's normal effects of motion, there is also an inertial force acting on the body at a right angle to its direction of motion. This force causes a body that would normally move in a straight line to follow a curved route. On Earth, the Coriolis force controls the overall wind directions and is also in charge of hurricane and tornado spinning.

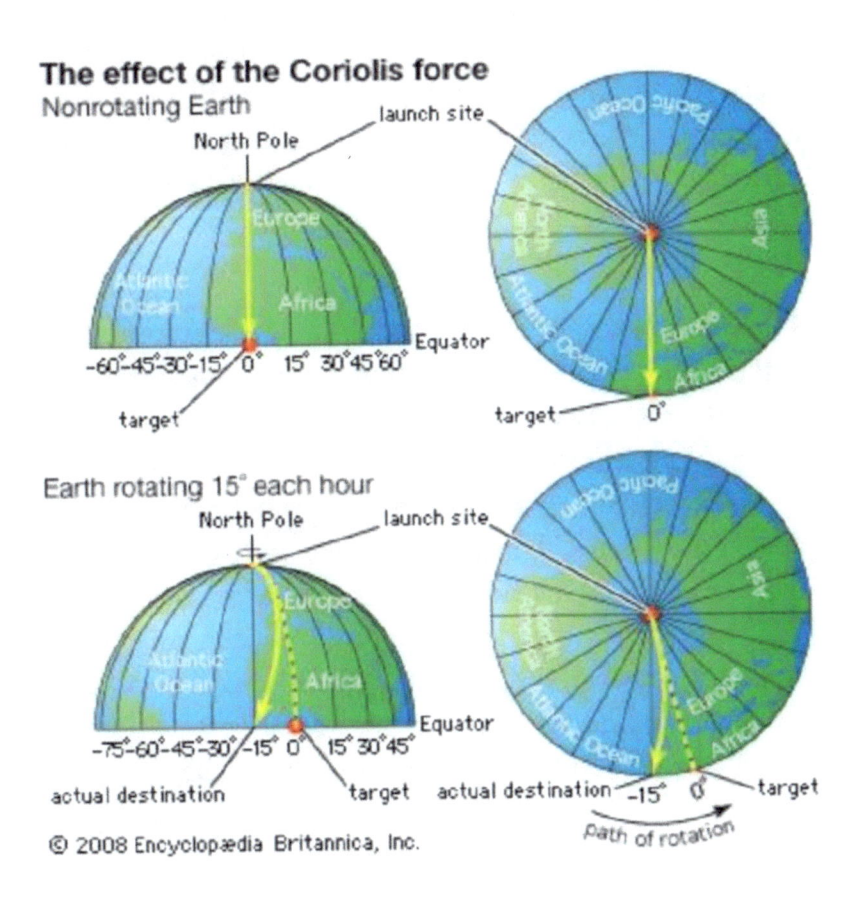

The effect of the Coriolis force

Nonrotating Earth

Earth rotating 15° each hour

© 2008 Encyclopædia Britannica, Inc.

44

The issue with the example from the Encyclopedia Britannica above is that there is no basis for it in reality. There is definitely a Coriolis Force. The Coriolis effect would be evident if the planet really were a spinning globe. The issue is that the earth does not experience the Coriolis effect. The earth is thus not rotating. The alleged Coriolis influence of the rotating planet has nothing to do with the varied rotational axes of storms in northern and southern latitudes.

The Coriolis effect is not taken into consideration while determining the flight trajectories of north and southbound aircraft. The Coriolis effect, for instance, would cause a plane flying from Buffalo, New York to Miami, Florida to veer off course in a westerly direction as the aircraft approaches the wider circumference of the earth at the latitude of Miami, Florida, assuming the heliocentric model with the earth traveling at more than 1,000 mph at the equator. However, the airplane actually makes it to Miami on schedule and without the pilot needing to make any Coriolis effect adjustments because of the earth's rotation.

In fact, landing an airplane on a runway would be extremely difficult if there was a Coriolis force. It would be difficult to line up the plane for a landing since a runway that runs north and south would be careening across the path of the aircraft at a speed of about 1,000 miles per hour.

There is a true Coriolis effect for rotating objects. For the spinning world to appear genuine, modern scientists must propagate the notion that the Coriolis effect is evident on the planet. For "scientists," the absence of the Coriolis effect on earth poses a significant challenge. They lie to get rid of that small issue. When a Coriolis effect is absent, they assert that one exists. The contemporary explanation for the Coriolis effect, which is intended to express itself on earth but is really entirely missing, is illustrated in the passage that follows from National Geographic.

Imagine that you are at the Equator and that your friend is in the center of North America. You want to toss a ball to them. Because your companion is going more slowly and hasn't caught up, the ball will seem to land to the right of him if you toss it straight.
Consider for a moment that you are standing at the North Pole. The ball will once more seem to land to the right of your friend when you throw it to him. However, this time it is because he is traveling more quickly than you are and has passed the ball.

The Coriolis effect is responsible for this apparent deflection.

The Coriolis effect affects rapidly moving things, like rockets and airplanes. When planning lengthy trips, pilots must account for the Earth's rotation. This implies that even though the airports are located precisely across the continent from one another, most aircraft do not fly in straight lines. For instance, the distance between Portland, Maine, and Portland, Oregon, is quite long and largely straight. A plane from Portland, Oregon, could not, however, fly directly to Portland, Maine, and land there. The Coriolis effect appears to curve to the right, toward the south, as one flies east. If the Oregonian pilot had flown the plane straight, it would have landed close to New York or Pennsylvania.

For identical reasons, military aircraft and missile-control equipment must compute the Coriolis effect. An air strike might destroy innocent persons and civilian structures while completely missing its intended target. ...

Movement on spinning objects is governed by the Coriolis force. It is based on two factors: the object's mass and rate of rotation. The Coriolis force is parallel to the axis of the object. From west to east, the Earth rotates on its axis. Therefore, the Coriolis force moves in a north-south direction. At the Equator, the Coriolis force is zero.

Although there is no physical force involved, the Coriolis force is useful in mathematical problems. Instead, it is simply the case that an item in the air is travelling faster than the ground.

The National Geographic is but one illustration of a widespread deceit. At the equator, the planet is intended to spin at a speed of around 1,000 miles per hour. The planet spins less quickly at higher and lower latitudes because the circumference of a ball is smaller north and south of the equator. Portland, Oregon is 45 degrees north of the equator, and it is said that the planet spins at a speed of about 700 miles per hour at that latitude. At 44 degrees North latitude, Portland, Maine, the planet spins only little faster than 700 miles per hour. If the pilot only attempted to fly the aircraft straight and level toward Portland, Maine, the Coriolis effect would theoretically place the aircraft in New York. Simply said, that is untrue. The pilot determines his course for Portland, Maine, taking into consideration solely the wind. Because the earth is not spinning and therefore there is no Coriolis effect to calculate, the pilot makes no allowance for one.

Military aircraft and missile-control technologies, according to The National Geographic, "must compute the Coriolis effect." For the sole reason that there is no authority, The National Geographic makes no reference to one in support of their claim. Because it is untrue, there is no authority. The National Geographic is only inventing information to deceive the uninformed audience into thinking that the world is rotating at a phenomenal rate.

According to the Coriolis effect theory, the eastbound aircraft in the National Geographic example would be able to maintain its speed while traveling at the same rate as the allegedly spinning earth because, upon takeoff, the aircraft would be accelerating above the 700 mph runway speed in Portland, Oregon. That reasoning has a flaw in that it presumes that the runway is aligned due east and that the jet is taking off in a due east direction.

The Coriolis effect is thought to be based on the earth's spin and the fact that once things are in motion over the spinning earth, they move independently of the earth's spin. If the Coriolis effect existed on earth, it would seriously affect air travel. A plane would never reach Portland, Maine if it took off from an airport in Portland, Oregon on a North/South runway and turned east to go there. This is due to the fact that after the aircraft reaches cruising altitude, it would be moving at a speed of around 560 miles per hour. However, beneath the plane, the earth would be spinning at a speed of 700 mph. The pace at which the planet spins would prevent the airplane from ever catching up. At a pace of 140 miles per hour, the plane would be continually dropping altitude above the ground. The aircraft would essentially be flying over the earth in reverse.

The precise reverse is what is discovered. In reality, a trip from Portland, Oregon, to Portland, Maine, heading eastward, would take less time than a flight from Portland, Maine, to Portland, Oregon, traveling westward. The cause is unrelated to how fast the planet is rotating. The jet stream, which is made up of high-altitude, high-velocity eastward winds, carries aircraft along while enabling eastward flights to travel at a quicker ground speed. Between 60 and over 250 miles per hour of wind may be found in the jet stream. Flights headed west have difficulties due to the same jet stream. A British Airways Boeing 777-200 airplane was said to have reached ground speeds of more than 745 miles per hour in January 2015 while traveling in the 250-mile-per-hour westbound jet stream. At sea level, the speed of sound is 760 miles per hour.

The earth is said to be immobile according to David Wardlaw Scott's explanation of experiments performed in England at the beginning of the 20th century using a cannon that revealed no Coriolis action at all. A gun was exactly vertically positioned on the ground and fastened securely during the testing. The cannon was let loose. The cannon ball was in the air for a total of 28 seconds after ascending vertically for 14 seconds and then returning to earth for another 14 seconds. The cannon ball should land about 5 miles to the west of the cannon if the earth were moving eastward at 600 miles per hour at the latitude in England. That, however, did not take place. Generally, the cannon ball missed the cannon by no more than two feet. The cannon ball did, in a few cases, really come back to the cannon's mouth

The Moon's Cold Light

We don't have to rely on Rowbotham's or the leading astronomers of his time. The eclipse is probably likely produced by a black, non-luminous body passing across the moon, as evidenced by the very nature of the moonlight. The fact that the moon does not turn black in the midst of the eclipse, as would be anticipated if sunlight were being blocked by a spherical earth, disproves the theory that an eclipse is caused by an earth shadow. In actuality, the moon seems to glow blood red during the peak of the eclipse. The moon takes on a dark crimson tint that resembles red-hot copper. In the middle of what should be the black umbra of the earth's shadow, the moon is really shining crimson.

By itself, this proves that the moon is a separate source of light rather than a mirror reflecting sunlight. That supports the account of God in Genesis. The sun and moon are both referred to as "lights" by God. God referred to the moon as a light in and of itself, not as a light reflector.

God then said, "Let there be lights in the firmament of the heaven to separate the day from the night; let them be for signs; let them be for seasons; let them be for days; and let them be for years; and let them be for lights in the firmament of the heaven to give light upon the earth," and it was so. And God created the stars in addition to two great lights, the larger light to rule the day and the lesser light to control the night. God placed them in the firmament of the heavens to provide light for the world, to govern day and night, and to separate light from darkness after determining that it was good.(Genesis 1:14-18)

Jesus was clear when he said that the moon does not act as a mirror; rather, it emits its own light. But in those times, following that affliction, the sun will be obscured and the moon won't shine. (Mark 13:24) The implication of the sun's darkness is that the moon won't be able to shine.
That clearly indicates that the moon often emits light on its own. See also Isaiah 13:10, Ezekiel 32:7, and Matthew 24:29.

True scientific observation demonstrates the Bible's veracity. It is demonstrable that moonlight is not a reflection of sunlight. In actuality, the moon emits light on its own. Therefore, the only explanation for an eclipse is the presence of a non-luminous mass in front of the moon, which is self-luminous. The idea that the moon reflects sunlight is completely refuted by Rowbotham's explanation of the special quality of moonlight.

A reflector won't emit heat or cold when they are supplied to it, respectively. Red light, not blue or yellow, will be sent back if a red light is received. When a musical instrument plays the note C, a reflector will not return the notes D or G, but the exact same tone, just with a slight degree or intensity difference.

If the moon is only a reflection of solar light, she is unable to emit or cast any other light than that which she initially receives from the sun. The light could not reasonably differ in any way other than intensity or amount; there could be no variation in the light's quality or character.

The only way to determine if the moon is a reflector is to check to see if the light we receive from her has the same or a different character from the light we receive from the sun.
First, the sun's light typically has an oppressive, violent, semi-golden, pyro-phosphorescent quality; in contrast, the moon's light is pale, silvery, and peaceful, and when it shines brightest, it is calm and non-pyrotic.

Second, the sun's light is warm, drying, preservative, or antiseptic; when exposed to it, animal and vegetable things quickly dry up, coagulate, contract, and lose their propensity to rot and become rotten. Therefore, due to prolonged exposure to sunshine, grapes and other fruits become firm, partly candied, and preserved, as seen in raisins, prunes, dates, and common grocery store currants. As a result, fish and flesh lose their gaseous and other volatile constituents through similar exposure, and through the coagulation of their albuminous and other compounds, they become firm and dry and less likely to decompose. In this way, a variety of fish and flesh that are well-known to travelers are preserved for use.

Animal and nitrogenous plant materials exposed to the moon's moist, chilly, and strongly septic light quickly begin to putrefy. Even living things suffer ill effects from prolonged exposure to the moon's light. On board ships traveling through tropical areas, written or printed signs are sometimes posted preventing people from sleeping on decks exposed to full moonlight. This is because experience has shown that such exposure frequently results in harmful effects.

Can Rowbotham's statements that moonlight and sunlight have distinct properties be backed up? It has really been demonstrated.

In the "Lancet" (Medical Journal), for March 14th, 1856, details are given of many studies that indicated that the moon's rays, when concentrated, really dropped the temperature upon a thermometer by more than eight degrees, according to Rowbotham. The temperature at the location where the amplified beam is cast rises as a result of the sun's rays being magnified, as is well known. Therefore, the warm light from the sun cannot be reflected by the moon due to its frigid amplified light.

The outcomes of the Lancet trials have been verified by experiments conducted as late as 2016. The frigid temperature induced by the moonlight was proved by magnifying the cold moonlight. One researcher utilized a Fresnel lens with high magnification (although he did not indicate its magnification strength). He noticed a temperature difference of about 8.1 degrees Celsius (14.6 degrees Fahrenheit) between the surface where the magnified moonlight was cast, which was 4 degrees Celsius (39.2 degrees Fahrenheit), and the nearby surface area that was shaded from the moonlight, which was 12.1 degrees Celsius (53.8 degrees Fahrenheit).

Is it essential to magnify the moonlight to determine how chilly it is? Not at all. Researchers have discovered that, even without the moonlight being magnified, items in the moonlight are noticeably cooler than those that are shielded from it.

A white ceramic plate put in moonlight (44.2°F) and a white ceramic plate shielded from the moonlight (51.4°F) had an about 7.2°F difference in temperature, according to one experiment. Another investigator calculated a difference of almost 2 degrees between the area of a metal ramp that was exposed to moonlight (54 degrees) and the area that was sheltered from the moonlight (56 degrees). Another tester discovered that a wallet placed on the ground in the moonlight was 4 degrees Fahrenheit cooler than a wallet left in the shadow from the moonlight.

Using a dual laser infrared thermometer, the author has found that material under moonlight is, in fact, cooler than similar material that is shielded from the moonlight. Depending on the substance, the temperature differential between material in moonlight and material shielded from the moonlight varied from 2 to 6 degrees Fahrenheit cooler. Of course, as was to be expected, the situation was completely reversed for material exposed to sunshine compared to material under shade. According to this source, materials exposed to sunshine were found to be between 20 and 25 degrees warmer than comparable materials that were shaded.

Moonlight differs from sunshine in that it is chilly, which makes it special (which is warm). That demonstrates that moonlight is not the sun's reflected light. If it were, the light that was reflected would need to be warm. As we've seen, an item in sunshine is warmer than an object shielded from sunlight, and an object exposed to moonlight makes it colder than an object covered from the moonlight by shadow. The two lights also differ in another way. While the heat from the sun is magnified when sunlight is present, the opposite is true of the cold when moonlight is present. Even when magnified tremendously, moonlight does not significantly lower temperature. Contrary to great amplification of sunlight, which results in a considerable temperature increase Dr. Henry Noad describes the outcomes of tests using magnified moonlight in his Chemistry Lectures:

Even when focused by the strongest blazing glass, the moonlight is unable to raise the temperature of even the most sensitive thermometer. Using a blazing glass that was 35 inches in diameter, M. De La Hire captured the full moon's rays while they were on the meridian and caused them to fall on the bulb of a sensitive air-thermometer. Although the moon beams were magnified 300 times by this glass, no effect was felt.

Professor Forbes used a lens with a 30 inch diameter, a 41 inch focal distance, and a power of concentration more than 6000 to focus the light from the moon. This lens cast the moon's image on the tip of a generous thermopile vividly. The moon was just 18 hours beyond full and less than two hours from the meridian. Even though the measurements were done in an ordinary way and the moon's light was concentrated 3000 times, not even the least thermal effect materialized (assuming that half of the rays were reflected, diffused, and absorbed).

Freemasonry and NASA

According to Thomas Africa, Nicolaus Copernicus (1473–1543) was not the pioneer of heliocentric astronomy but rather a restorer of Pythagoras of Samos' heliocentric kind of theory (570-495 B.C.). According to others, Pythagoras' system wasn't entirely heliocentric; rather, it was a system in which the sun and other planets orbited an unseen fire in the center. However, because Pythagoras' theory was the first to propose that the planets move in a circular orbit, he has been acknowledged as the genuine inventor of the heliocentric system by early scientists. Johannes Kepler did in fact (1571-1630 A.D.). the "grandfather of all Copernicans" referred to Pythagoras. In his own words, Copernicus argued that his approach was not novel but rather a rediscovery of Pythagoras's long-forgotten philosophy. The papal decree of 1616, in the opinion of Galileo Galilei (1564–1642), suppressed the "Pythagorean idea of the mobility of the earth.

Copernicus also "borrowed" from Aristarchus of Samos' (310–230 B.C.) beliefs that the earth revolved around the sun. Pythagoras' contributions were readily acknowledged by Copernicus, but for some reason he chose to keep his familiarity with Aristarchus' works a secret. According to Thomas Africa, Copernicus was aware of Aristarchus' heliocentrism but kept it a secret, eventually deleting his one fleeting mention of it out of either vanity, "Pythagorean" scruples, or both.

The concept of spherical planets orbiting a central fire in a circular fashion was initially put forth by Pythagoras. He allegedly created a counter-earth to get to 10 circling planets (including the sun). According to Jose Wudka, the additional counter-earth was created to account for eclipses and because ancient thinkers revered the number 10 as sacrosanct. The 10 circling spheres coincidentally correspond to the 10 spherical sefirot of the Jewish Kabbalistic divinity Ein Sof.

Mason Master In his book, Defense of Masonry, Dr. James Anderson, the founder of the London Masonic Lodge, claimed that Pythagoras was a forebear of Freemasonry. Ancient Masonic documents suggest that the root of Freemasonry lies in Pythagorean ideas, according to Master Mason William Hutchinson, who wrote about this in his book Spirit of Masonry. In his Illustrations of Masonry, William Preston, another Master Mason, claims that

Pythagoras was initiated into the profoundly enigmatic Masonic doctrines, which he then disseminated to the nations through which he journeyed. Regarding Pythagoras and his Masonic ties, Albert Mackey writes the following in the Encyclopedia of Freemasonry:

Pythagoras built his renowned school in Crotona, a Dorian Colony in the south of Italy, around 529 B.C., which is quite similar to the later-adopted Freemasonry. The candidate's past life and character were rigorously examined before they were granted access to the school's privileges. During the preliminary initiation, secrecy was commanded by an oath, and he was forced to endure the most trying tests of his fortitude and self-control. The Crotona School's way of life was similar to that of contemporary Communists.

Some of the philosophies of the Kabbalah are classified as Pythagorean in the Jewish Encyclopedia. The Kabbalah and Pythagorean occultism both derived from Babylonian mysticism. According to the Jewish Encyclopedia, Gnosticism is of Chaldean (i.e., Babylonian) origin and has Jewish characteristics. That suggests that Jews in Babylon are where Gnosticism first emerged. The mysticism that the Jews adopted when they were captive in Babylon is memorialized in the Kabbalah.

The Jewish Encyclopedia appears to use the adjective "Pythagorean" to describe the type of beliefs found in the Kabbalah. They have characteristics with Pythagorean beliefs. Considering that Pythagoras was born some 27 years before the Jews were first taken into captivity in Babylon, in 597 B.C., this cannot possibly suggest that Pythagoras was the source of the Kabbalah. In or about 538 B.C., the Jews were freed from their Babylonian captivity.

A Greek, Pythagoras was. The first time he went to Babylon was around 525 B.C. According to legend, Pythagoras was imprisoned there for five years. The Syrian philosopher Iamblichus, who was born in or around 250 A.D., claims that Pythagoras "was brought by the supporters of Cambyses as a prisoner of war.
While he was there, he was happy to interact with the Magoi, learn about their very mystical deity worship, and get instruction in their holy ceremonies. He also attained the pinnacle of excellence in music, arithmetic, and other Babylonian mathematical disciplines.

None of the tenets of the Kabbalah could have originated from Pythagoras. Because by the time Pythagoras arrived on the scene, the Jews had already been kept as slaves in Babylon, exposed to the occult mysteries of Babylon, and freed from their slavery. By the time Pythagoras arrived in Babylon, there were undoubtedly many Jewish mystics still there. It's possible that Pythagoras received his initiation into the secrets of what is now known as the Kabbalah during that time.

The Jewish Encyclopedia appears to be more paying respect to Pythagoras by referring to certain of the Kabbalistic concepts as Pythagorean when describing the nature of specific doctrines contained in the Kabbalah. In fact, when asked to specify where the mysticism in the Kabbalah originated, the Jewish Encyclopedia made it clear that gnosticism is of Jewish nature and has "Chaldean [i.e., Babylonian] roots.

There may have been some synergism in the exchanges between Pythagoras and the Kabbalistic Jews, but it does not change the fact that the occult Babylonianism serves as the foundation for both Pythagoras' and Jewish mystics' philosophical systems. The fact that the Babylonians were aware of the Pythagorean theorem, for which Pythagoras is renowned, a thousand years before Pythagoras attests to the Babylonian roots of his philosophy. Pythagoras was a Kabbalist of the greatest rank, according to S. Pancoast, who treated the notorious occult theosophist H.P. Blavatsky. He goes on to say that Pythagoras was aware of the Kabbalistic roots of the Masonic emblems.

According to Pancoast, Pythagoras was introduced to the mysteries of the Kabbalah during his initiation, which introduced him to the heliocentric theory.

Pythagoras believed that the Sun is the center of the solar system and that all of the planets revolve around it. He also believed that the stars are similar to our Sun and are each the system's center. Pythagoras also believed that the planets are inhabited and that they and the earth are constantly rotating in a predictable pattern.

German humanist Johannes Reuchlin served as the Chancellor of Germany's political advisor from 1455 until 1522 A.D. He was a specialist in the languages and customs of antiquity and a classicist (Latin, Greek, and Hebrew). Reuchlin had ties to della Mirandola and other leaders of the Platonic Academia. Pythagoras, the father of philosophy, did not, however, get such teachings from the Greeks; rather, he learned them from the Jews, according to Reuchlin, who also proved that Pythagoras acquired his philosophy through the Jewish Kabbalah. He must thus be referred to as "a Kabbalist," and he was the first to change the word "Kabbalah," which was unfamiliar to the Greeks, into the Greek name philosophy. The philosophy of Pythagoras sprang from the limitless depths of the Kabbalah.

The Jewish Kabbalah served as the foundation for the religion of freemasonry, which may be traced back to Pythagoras. Masonry is a pursuit for Light, according to Albert Pike in Morals and Dogma. You see, that takes us straight back to the Kabalah. Albert Pike's assertion is supported by Albert Mackey's official Encyclopedia of Freemasonry. According to Mackey, the Kabbalah, which is "the mystical philosophy or theosophy of the Jews," is closely related to the symbolic science of Freemasonry.

In composing his official Morals and Dogma of the Ancient and Accepted Scottish Rite, Albert Pike used what source? Freemasonry: An Interpretation, a book on his discoveries, was written by Martin L. Wagner after he performed an extensive and unbiased investigation of Freemasonry.

For the purpose of elucidating and showing Freemasonry, Wagner claims that Albert Pike "drew heavily from the teachings of Eliphas Levi, the Abbe Constant, a famous Kabbalist, and whom Buck thinks to know more of the esoteric knowledge than anybody since the days of the old initiates." In its purest form, the Kabbalah is witchcraft. The Kabbalah has elements of magic and occult mysticism throughout. The Kabbalah incorporates a lot of dark magic, sorcery, and calling upon the devil's powers.

The hidden use of purportedly Gentile countries and organizations to further Jewish Zionist goals while concealing Jewish control over such institutions is one of the primary revelations made in the Protocols of the Learned Elders of Zion. The Learned Elders of Zion claim in the Protocols that they have concealed their participation in the scheme for a "new world order" by using Masonry as a front. Protocol 4's second paragraph states:

Who or what has the power to unsettle an unseen force? And precisely this is the nature of our energy. Gentile masonry blindly acts as a cover for us and our goals, but our force's strategy and even its permanent location are unknown to the general public.

The Zionist Jews have the ideal cover thanks to the Freemasonry's Gentile façade. The "Christian" Zionist movement is under the same covert Jewish influence, as is evident. The foundation of freemasonry is judaism. It is a Gentile front for Jewish mysticism, which has Jewish theosophical roots in its history, education, and positions of authority.

Masonic lodges are supported and governed by Zionists. They employ the lodges as vital, covert intelligence gathering organizations and power centers. In accordance with Protocol 15's paragraphs four and five, "[W]e shall create and multiply free masonic lodges throughout the world, incorporating into them all who may become or who are prominent in public activity, for in these lodges we shall find our chief intelligence office and means of influence." All of these lodges will be unified under one central government, known only to us and completely unknown to anybody else, that will be made up of our wise elders.

It is only reasonable that we, and no one else, should be in charge of masonic activities since, unlike the Goyim, we know where we are going and what the end purpose of each action is.

Zionist Jews recruit their front men from the Gentile community through Freemasonry lodges. Freemasonry's alleged Gentile origins are really a facade; it is fully derived from the Jewish Cabala. Protocol 11's paragraphs four and seven show how the Gentile front of freemasonry is used to further Zionist goals.

We are the wolves, and the Goyim are a herd of sheep. And you are aware of what occurs when the wolves capture the flock? ... Then why did we create this entire policy and ingrain it into the brains of the GOEY without allowing them the opportunity to consider its underlying meaning? Indeed, for what purpose if not to circumvent the fact that the direct route is impassable to our dispersed tribe? It is this that has served as the foundation for our organization of secret masonry, which is unknown to and pursues goals that are not even somewhat suspected by these "goy" livestock, drawn by us into the "show" army of masonic lodges in order to cast scorn in the eyes of their compatriots.

Wagner's research on Freemasonry supports the claims made in the Protocols that Freemasonry had a Jewish origin.
In his book, Wagner cites Masonic sources who claim that "Masonry in its purity, deriving as it is from the old Hebrew Kabbala as a component of the great universal wisdom religion of remotest antiquity.

According to Wagner's analysis, Freemasonry owes a great deal of debt to the Kabbalah for its philosophical beliefs, its techniques for understanding the Bible, its theories on emanations, its artistic language, its cosmological viewpoints, and its veils and glyphs. It is, in a way, the Kabbalah continued under a different name and persona.

The renowned Rabbi Isaac Wise provides more evidence of the Judaic roots of Freemasonry. Wise affirms that Freemasonry's ostensibly Gentile character is really a facade: "Freemasonry is a Jewish establishment, whose history, credentials, positions of authority, codes of conduct, and explanations are Jewish from beginning to finish. Masonry is founded on Judaism, according to the Jewish Tribune of New York, published on October 28, 1927. What remains of the Masonic Ritual if the Jewish teachings are removed?

Michael Hoffman came to the following conclusion: "The Freemasons and other occult works of iniquity take their views from these [Cabalistic and Talmudic] recondite theories of Judaism. Freemasonry is referred to by Henry Makow as "Judaism for Gentiles." It is "a mechanism for the Cabalistic Jewish elite to recruit Gentiles into their scheme," according to Makow.

What are the spiritual principles that Freemasonry is built upon and which come from the Kabbalah? It consists of worshiping Lucifer. Freemasonry's theological pope, Albert Pike, explains:

We must tell a group of people that while we worship a God, it is a God that is adored without superstition. The Masonic Religion should be upheld by all of us initiates of the high degrees in the utmost purity of the Luciferian Doctrine, we say to you, Sovereign Grand Inspectors General, so that you may repeat it to the Brethren of the 32nd, 31st, and 30th degrees. Would Adonay and his priests defame Lucifer if he weren't God, given that his actions demonstrate his brutality, perfidy, hate of people, barbarism, and aversion to science? Unfortunately, Adonay and Lucifer are both deities. The ultimate can only exist as two gods: darkness being essential for the statue and the brake to the locomotive. According to the everlasting rule, there cannot be light without shade, beauty without ugliness, or white without black. As a result, the idea of Satanism is heresy, and the belief in Lucifer as the equal of Adonay is the real and pure intellectual religion. However, Lucifer, the God of Light and God of Good, is engaged in a battle for mankind against Adonay, the God of Darkness and Evil.

The Hebrew term for God, Adonay, is used throughout the Old Testament and is rendered as "Lord" in the English version of the Bible. Pike insults God by referring to him as "the God of Darkness and Evil." The "God of Good," as Pike refers to Lucifer. Pike acknowledges that the Masonic deity of light is Lucifer. And there's no reason to be surprised because

Satan has changed into an angel of light. (See 2 Corinthians 11:14). Pike honors the deity of Freemasonry in Morals and Dogma, his authoritative work that continues to serve as the theological bible of Masonry: "Lucifer, the Bringer of Light! A strange and enigmatic moniker for the Spirit of Darkness! The Son of the Morning is Lucifer. Is it he who carries the light, which with its insufferable splendors blinds weak, sensual, or self-centered Souls? Not a chance! In actuality, the name Lucifer means "light bearer." The aspirant keeps looking for greater illumination during the Masonic initiation rites. The applicant will be told that the light he seeks is found in the light bearer, Lucifer, who is the deity of Freemasonry, if he or she advances to the highest degree of Freemasonry.

"[W]hen the Mason realizes that the secret to the warrior on the block is the appropriate application of the dynamo of living force, he has mastered the mystery of his Craft," says Manly P. Hall, a 33-year-old Freemason and highly regarded Masonic authority.

He has the boiling energies of Lucifer, but he must first demonstrate his mastery of energy manipulation in order to go forward and upward. The Kabbalah's central principle is heliocentrism. Heliocentrism is therefore essential to Freemasonry. In several countries throughout the world, Masonic lodges bear Copernicus' name. The Freemasons' Quarterly Review, published in 1843, praised Copernicus in the passage below.

By demonstrating that all heavenly body motions serve to advance the honor and glory of the Great Architect of the Universe, Copernicus and his predecessors in the study of the starry sky have rendered the practice of astrology obsolete.

The tight ties between Freemasonry and NASA are explained by the connection between heliocentrism and Freemasonry. For instance, James Edwin Webb, the NASA administrator from 1961 to 1968, belonged to the Freemasons. Kenneth S. Kleinknecht, a 33° Freemason who served as the manager of the Apollo Program Command and Service Modules, the deputy manager of the Gemini Program, and the manager of Project Mercury, published an essay in the November 1969 issue of the Masonic magazine The New Age. By the way, C. Fred Kleinknecht, 33°, Sovereign Grand Commander, The Supreme Council (Mother Council of the World), Southern Jurisdiction, USA, Washington, is the brother of Kenneth S. Kleinknecht. Kenneth Kleinknecht wrote in an essay in The New Age:

Take note of how many of the astronauts, including Edwin E. Aldrin Jr., L. Gordon Cooper Jr., Donn F. Eisele, Walter M. Schirra, Thomas P. Stafford, Edgar D. Mitchell, and Paul J. Weitz, are Brother Masons themselves. Virgil I. "Gus" Grissom was a Mason as well before he tragically perished on January 27, 1967, in a flash fire at Cape Kennedy. During his historic Gemini V space mission in August 1965, astronaut Gordon Cooper carried a Scottish Rite flag and an official jewel from the Thirty-third Degree.

Masonry has already entered the space era as evidenced by the lunar plaque, the Masonic flag and ensign, and the Masonic astronauts themselves. Can we seriously question Freemasonry's spiritual significance in the current day when even its tangible representations have made significant recent forays into the boundless reaches of space?

The Grand Lodge of Texas, A.F. & A.M., has published the following justification for chartering Tranquility Lodge No. 2000, a Masonic lodge on the moon, on the internet.

Two American astronauts made their first lunar landing on the Mare Tranquilitatis, or "Sea of Tranquility," on July 20, 1969. Brother Edwin Eugene (Buzz) Aldrin, Jr., a member of Clear Lake Lodge No. 1417, AF&AM, Seabrook, Texas, was one of those heroic men. Brother Aldrin traveled with the **SPECIAL DEPUTATION** of then-Grand Master J. Guy Smith, establishing and appointing Brother Aldrin as the Grand Master's Special Deputy, giving him full authority to act in the Grand Master's place of business and allowing him to claim Masonic Territorial Jurisdiction for The Most Worshipful Grand Lodge of Texas, Ancient Free and Accepted Masons, on the Moon. The Special Deputation also instructed him to make a proper accounting of his Brother Aldrin attested that he was the one who brought the Special Deputation to the Moon on July 20, 1969. 363
(Original emphasis in bold)

Figure: Freemason Buzz Aldrin (right) with Luther A. Smith, the Masonic Sovereign Grand Commander, holding the Masonic flag Aldrin took with him when he allegedly landed on the moon.

Alex Jones spoke with Buzz Aldrin in 2009.
The following conversation took place between Alex Jones and Buzz Aldrin during the interview:
"Mr. Aldrin," says Alex Jones I've had this question for you forever. We saw images of the little Masonic flag on the moon, some of the mission names, and numerology; is there any significance to any of that? And what exactly is the Masonic influence? What is the Masonic effect on NASA? We are aware that the country's founding has Masonic influences.

Aldrin: "Zero, as far as I can tell. In Texas, I had some Masonic brethren who wanted me to send some sort of Masonic symbol to the moon as a gesture. I'm not sure what that would be, but I informed them it was not within my power to send something like that.

In a letter he addressed on September 19, 1969, three days after meeting with the Scottish Rite of Freemasonry's leaders in the House of the Temple in Washington, D.C., Aldrin acknowledged giving them the identical flag he had claimed to have carried to the moon in an Alex Jones interview in 2009.

Additionally, a picture of Aldrin presenting the "Masonic insignia" he purportedly brought with him to the moon and back was included in a story in the December 1969 edition of the Masonic magazine New Age, the official publication of the Scottish Rite Southern Jurisdiction.

Why would Aldrin make a false statement regarding NASA's ties to the Masons? Because it would expose the true influence and power behind NASA. The hidden agenda would be revealed if that thread were pulled. Didn't Aldrin know that in a letter he addressed to the leaders of the Scottish Rite of Freemasonry on September 19, 1969, he claimed to have brought a Masonic flag to the moon? He was aware of the letter , and the Masonic fraternity was aware of his assertion that he owned a Masonic flag on the moon. He was probably unsure of which of the numerous conversations he had with Masonry were public because so many within the organization are private and sealed by blood pledges. He simply could not distinguish in his mind between the few public Masonic communications and the secret Masonic communications because his life is so compartmentalized between his public facade, which is largely based on deception, and his private Masonic communications, which are for the most part secret. Aldrin didn't have much time to consider his response to Alex Jones' surprise question, so he just went with his go-to tactic: lying.

There has been (and presumably now is) a dominant group at NASA made up of a statistically improbable majority of Masonic astronauts and officials. In actuality, the list of Masonic astronauts that Kleinknecht includes in his New Age piece is far from exhaustive. For instance, Freemasons James Irwin (Apollo 15), Fred Haise (Apollo 13), and John Glenn Jr. (Mercury 6) are all absent from Keinknecht's list of notable astronaut Freemasons.

Freemasons are quite proud of their association with NASA. Aldrin's denial of any Masonic ties to NASA raises the possibility that the Masonic brotherhood should be the only group to know about the relationship. The Masonic medallion shown below was produced to mark the tenth anniversary of the purported Apollo 11 lunar landings.

The Phoenix Freemason Museum website gives the medallion the following description:

The 10th anniversary of our flags on the moon was celebrated with the creation of this medallion in 1979. Many people were unaware that astronaut and brother Neil Armstrong brought two flags with him on his Apollo mission to the moon. The American flag was one, and the Scottish Rite's Southern Jurisdiction's banner, which had the double-headed eagle insignia, was the other. This flag is currently housed in the Scottish Rite Museum's museum collection in the House of the Temple. It has a 1 3/4 inch diameter and is an amazing three-dimensional medallion. The Medallic Art Co. of Dansbury, Connecticut created it.

In his book, Dark Mission: The Secret History of NASA, Richard Hoagland, a former science advisor to CBS News during the Apollo program from 1968 to 1971, claims that Freemasons are in charge of NASA and that the agency has always had an occult underbelly that has been carefully hidden from the general public. Although Hoagland's claim is accurate, Bill Kaysing has grounds to think that Hoagland is a NASA shill who is attempting to deflect attention from the fact that NASA never sent a man to the moon. From 1956 through 1963, Kaysing oversaw the technical papers for Rocketdyne's whole propulsion laboratory, which served as a testing ground for huge liquid rocket engines. The subsidiary of North American Aviation and subsequently Rockwell International that produced the Saturn V rockets utilized in the NASA Apollo Missions was known as Rocketdyne. Kaysing had access to papers relevant to the Mercury, Gemini, and Apollo programs when he was employed by Rocketdyne thanks to his top secret clearance. The Rocketdyne experience and subsequent study convinced Kaysing that the Apollo lunar landings were a total fabrication. We Never Went to the Moon: America's Thirty Billion Dollar Swindle, a book that was initially released in 1976, contains the arguments he used to support his claims.

Kaysing's accusation against Hoagland seems to have some substance. Hoagland has attempted to spin the evidence of the obviously fabricated NASA images and movies, not as evidence that the moon landings never occurred, but rather as evidence that NASA is hiding the existence of space aliens. This is known as a limited hangout in the intelligence community and occurs when a shill seems to uncover a small portion of a conspiracy in order to divert attention from the reality and offer a credible (though less sinister) justification for the government's lie.

Hoagland claims that just a small percentage of the lunar images and videos are phony. He promotes the view that NASA traveled to the moon and that it only altered a small portion of the images to conceal signs of intelligent life. Hoagland remarked in a radio interview:

I'll tell you what my conclusion is. Although I believe there is a sizable Apollo conspiracy, I believe the correct conspiracy has been marketed to us in order to divert the attention of intelligent individuals like you who are asking insightful questions—something these people are masters at doing. The true question isn't whether we went to the moon, but rather, what did we discover there that they don't want you to know. I've discovered places where NASA allegedly altered the images to conceal some very fascinating stuff.

Hoagland overstates his points and presents false data. By exposing mistakes that have been planted by its own shills, the intelligence agency is able to undermine resistance to its programs. Verifying the information is important. It is for this reason that the author has supplied endnotes citing the sources of almost all of the information in this book.

Hoagland offers some verifiable facts (such as the occult Masonic influence in NASA and the phony images and films), but she turns their relevance away from moon landing hoaxes and toward extraterrestrial life. That serves to refute the claims of occult practices and phony images, and as a result, many people discount the idea that the moon landings were a hoax. Other individuals may find Hoagland's notion of NASA manufacturing movies and images in order to hide life on other planets to be a more logical and less sinister explanation for the agency's deceit, which helps to distract them from the reality that NASA did not travel to the moon. NASA exists to keep the public brainwashed into believing the Satanic deception of a rotating, circling planet where man is but a minor component of an endless, godless world. Hoagland advances that goal. The majority of those who have uncovered NASA's deceit appear to pay little attention to the purpose behind fabricating the moon landings. Many people attribute NASA's motivation to financial gain, however some also attribute it to prestige during the Cold War or as a diversion from the Vietnam War.

Certainly, NASA defrauded American taxpayers of billions of dollars. Those responsible for the moon landing hoax conspiracy undoubtedly made money off the money scam. The true goal was to instill a love of money in men's hearts so they would use it to rule and enslave the world, not to make a quick profit. The belief that money is the source of all evil. Timotheus 6:10 The purpose of the moon landing hoax was to encourage the love of money to grow. The soil must be tilled by the untruth that there is no God in order for the root of evil to spread far into it. Men need to be persuaded that there are no everlasting repercussions for lying, deceiving, and stealing in order to obtain money in order to build the love of money in their souls. Men must have this view in order for there to be no God who can administer sin punishment. This calls for keeping man in the dark about the fact that he was created by God, in his image, on a flat earth that sits at the center of his creation. The purpose of the moon expedition is to deceive humanity into thinking that there is no God and that there is an unending cosmos with just man existing on a small, spherical planet careening through space.

The key to comprehending the moon landing hoax is realizing that it is spiritual rather than scientific deceit. The use of symbols is crucial in witchcraft. To signify that it is controlled by the great serpent, Satan, the NASA emblem has a snake's forked tongue. And the big dragon, that ancient serpent known as the Devil and Satan, who deceives the entire world, was cast out into the soil, together with his angels. Chapter 12:9.

The Apollo name for NASA's lunar missions is noteworthy. The Greek sun deity is called Apollo. Apollo is frequently seen driving a chariot driven by horses as the sun is beaming behind him. You'll notice that Apollo is not depicted in the Apollo XIII symbol. Although there isn't a chariot, it appears that the horses are leaving the earth in the illustration. The horses might be seen as pulling the earth behind the chariot, suggesting that the earth itself is Apollo's chariot in light of the missing chariot. The god of this world is referred to as Satan in the Bible (2 Corinthians 4:3–4), who closes the eyes of the lost to the light of the gospel of Jesus Christ.

Official NASA Emblem for the Apollo XIII Mission

The angel of the bottomless pit, Apollo, is the same Apollyon that is mentioned in the book of Revelations. "And they had a king over them, who is the angel of the bottomless pit; and his name is Abaddon in the Hebrew tongue, but his name is Apollyon in the Greek tongue." (9:11 in Revelation) According to the Edinburgh Encyclopedia:

There can be little doubt that the Pythian Apollo is the same as the Hebrew gods Ob and Abaddon, which the Greeks interpreted literally as Apollyon, according to Hensius' analysis of this verse.

The allusion to Apollyon in Revelation 9:11 is a reference to Apollo, who is Satan, according to Wakeman Ryno in his book Amen: The God of the Amonians Or a Key to the Mansions in Heaven. "The Sun (Apollo) in the Sign of the Scorpion, the ruler of the bottomless pit, is the same as Satan, Belial, Lucifer, Abaddon, and Apollyon (parenthetical in original). Lucifer is the god of Freemasonry, thus it is not unexpected that NASA's moon missions were given the name Apollo in honor of the Masonic god (Lucifer). The world was largely brainwashed into accepting the falsehood of a rotating, spherical Earth circling the sun by the NASA Apollo missions.

The Satanic goals of NASA are summarized by Texe Marrs as having Masonic roots:
The concepts of Masonic alchemy and the mysticism of the ancient mystery religions served as the foundation for NASA's space program from the beginning. The antichrist king, as predicted by the prophet Daniel, will be a powerful world leader in the coming days, "And through his policies he shall cause cunning to thrive," Daniel said. It's craft, as in witchcraft! In the early stages of the American space program, the covert OSS/CIA initiative known as Operation Paperclip saw Nazi rocket experts like Werner Von Braun transported from war-torn Germany to America and given charge of the creation of spacecraft.

The newly established space organization was subsequently given up to the Freemasons, and magic and witchcraft were combined with the most recent scientific discoveries. Almost everything NASA does is infused with magic and alchemy. A second matrix contains NASA's true mission, which is concealed from the general public. While the mind-controlled and manipulated masses are driven into ever-increasing levels of altered consciousness, this process entails the construction of Satanic ritual magic that allows the Illuminati elite to obtain and accumulate power.

The Jewish Kabbalah is the parent of Freemasonry. Since Zionism controls Freemasonry at its highest levels, it follows that Zionism eventually controls NASA. Acronyms and terms have several meanings in witchcraft. Obviously, the National Aeronautics and Space Administration is referred to as NASA. The Hebrew term "nasa" also implies to hoist up, carry off, or glorify oneself. Additionally, the Hebrew term Nasa can be translated as "make to suffer sin."

Zionist influence at NASA is well-kept secret. However, there are signs that Zionism is influencing NASA. The Columbia Shuttle Mission STS-107 logo is one illustration. The Columbia Shuttle was supposedly destroyed on February 1, 2003, as it purportedly reentered the atmosphere during the tragic Space Shuttle Mission STS-107. It is appropriate etiquette for a memorial symbol to have both the flag of the host nation and the flag of a visiting astronaut. The Columbia Shuttle Mission STS-107 insignia stood noteworthy for its flagrant violation of convention, which consisted of simply showing the flag of the visiting astronaut's nation without doing the same for the host nation, which in this case was the United States. Which nation was the visiting astronaut from? He was an Israeli native. It was meant as a not-so-subtle sign of Israeli control over NASA and the US Government because the Israeli flag appeared on the Columbia Shuttle Mission STS-107 logo without the American flag as well.

The appearance of the Israeli flag and the absence of the American flag is omen of future events. In the end, Israel will stab the United States in the back. Israel will take advantage of the US as much as it can before doing that. In fact, after visiting Jonathan Pollard in jail, Israeli Prime Minister Benjamin Netanyahu was overheard by a CIA operative telling his followers that "[o]nce we squeeze all we can out of the United States, it may dry up and blow away." Jewish spy Pollard was exposed as an Israeli snoop on the United States. I failed in the Pollard issue, just as I failed in previous intelligence operations outside of enemy lines, Rafi Eitan, a Mossad spymaster, Fidel Castro counselor, and Israeli cabinet member, said to Yediot Aharonot in June 1997. According to Eitan, Israel views the United States as an adversary. Read 9/11-Enemies Foreign and Domestic by this author for additional details on Israel's hypocrisy.

For a very long period, there has been strong cooperation between **NASA** and the Israel Space Agency (**ISA**). As an illustration, **NASA** and **ISA** formally agreed to exchange technologies in 1986. Through that arrangement, Israel will receive a significant amount of very sensitive technology that was produced by **NASA** at American money.

It's interesting to note that according to the American-Israeli Cooperative Enterprise, academics from Ben-Gurion University joined a global initiative to map the world in October 1999 that was funded by **NASA**, the German space agency **DARA**, and the Italian space agency **ASI**. That mapping of the world endeavor is undoubtedly a part of the global campaign to deny the flat planet. Israel asserts that a refrigerator made in Israel was part of the allegedly successful **NASA** Curiosity Rover mission to Mars in November 2011. Israel is interested in **NASA**'s technology, and **NASA** is far too accommodating. That explains, for instance, why Israel chose to host the 2015 International Astronautical Conference (IAC), which took place in Jerusalem. Israel revealed another collaboration between **NASA** and the Israel Space Agency (ISA) at the conference, involving "combined missions, people and scientific data exchanges, ground-based research facilities.

Although the agreements with the US appear to be bilateral, they are actually one-sided. There is little doubt that Israel possesses cutting-edge technology, but it has very little to offer the United States. Israeli technology is almost entirely a gift from or a theft from another nation, mainly the United States. Israel, though, does exchange American technology with other like-minded nations. The creation of nuclear and other military weapons has long been a shared endeavor between Israel and communist China, albeit in secret. Indeed, one of the main routes for the transfer of American and other western technologies to communist China has been identified as Israel. Patriotic Americans would never sign deals with Israel that favor Israel exclusively in terms of technology exchange. Because Israel owns **NASA**, the biased agreements were approved by **NASA** authorities. Masonic **NASA** officials merely comply with Israel's demands. ISA is more of an intelligence organization than a true space agency. Its goal is to acquire as much American technology as it can. The technology sharing agreements between NASA and ISA are actually technology handover agreements, wherein Israel is given access to advanced U.S. technology. Read this author's book, Bloody Zion, for a thorough description of Israel's stranglehold over the United States Government.

The Flat Earth Map

The fact that the world is frequently shown as a globe since infancy makes it difficult for most people to comprehend the flat planet. Usually, a globe of the planet was the first scientific item shown to the students. In their earliest global history studies, they learned about Magellan's voyage across the world.
It will be simple to see how Magellan might have been considered to have circumnavigated the globe while in reality he just circumnavigated a plane once one understands how the flat earth actually exists.

The earth is shown to be flat in the illustration below, with the north pole in the middle of a circle. At every point along the plane's edge, the south is present. In reality, Antarctica encircles the whole earth's surface.

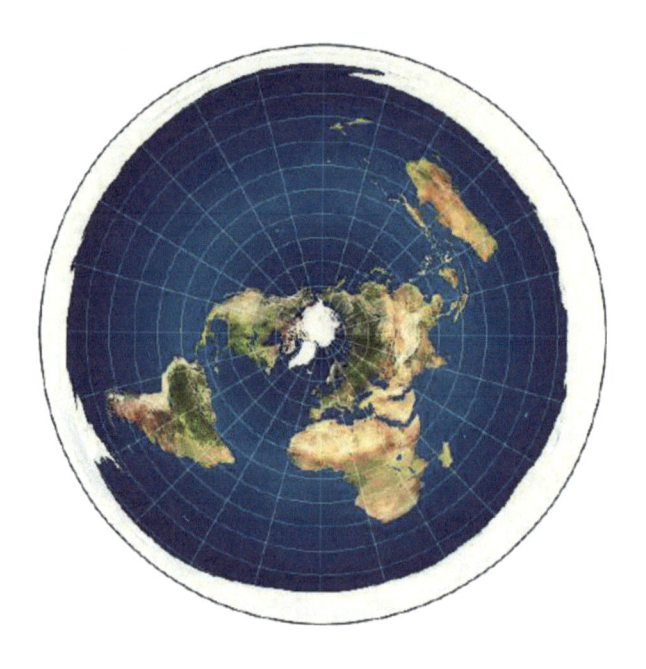

In actuality, the map seen above is referred to as a polar azimuthal equidistant map. The United States Geological Survey (USGS) asserts that azimuthal maps are reliable for depicting continents and seas. They are utilized for air and marine navigation because to their precision. In fact, according to the USGS, Micronesia's large-scale mapping and the National Atlas of the United States of AmericaTM both employ azimuthal equidistant maps. useful for displaying airline distances from the projection's center. The earth does not have to be a globe in order to be circumnavigable. Magellan traveled across the entire flat earth.

The edge of the flat earth is Antarctica. When explorers first arrive in Antarctica, they are confronted by a gigantic ice wall that is 1,000–2,000 feet thick, with 100–200 feet of that thickness rising above the ocean. 403 British Naval Officer and polar explorer Sir James Clark Ross was one of the first travelers to view the Antarctic ice wall. In front of him was a sheer ice cliff with a flat top that Ross estimated to be between 150 and 200 feet high and stretching in both directions as far as the eye could see. Famously, Ross described the ice wall in the Antarctic:

There was no question in my mind about how it would affect our future actions because we could just as easily sail through the Dover cliffs than try to go past such a mass. There was no way to imagine a mass of ice seeming more solid; not the slightest sign of any crack or split could be seen throughout its whole expanse, and the incredibly clear sky beyond it just served to further emphasize how far south it extended.

Antarctic Ice Wall

The azimuthal equidistant map is accurate in all directions and distances from the map's center, according to the USGS. However, according to the USGS, the map is only precise when the beginning point is in the middle of the map. Therefore, the USGS acknowledges that all distances and directions from the North Pole to anywhere on the map traveling south are accurate in the flat earth map shown above, but asserts that they are not always accurate when starting from any other point on the map other than the North Pole.

Look at how the flat earth map and the UN flag emblem below are similar. Antarctica is the sole omission from the UN's flag. The skeleton world government, which was created to subjugate humanity, has kept the flat earth secret out in the open.

Also take note of the UN Flag's logo's several components. There are precisely 33 divisions, which matches up perfectly with the Scottish Rite of Freemasonry's 33 degrees.

Where is the equator, one would ask? The Tropic of Capricorn, the Tropic of Cancer, and the Equator are all clearly marked on the Polar azimuthal equidistant map shown below.

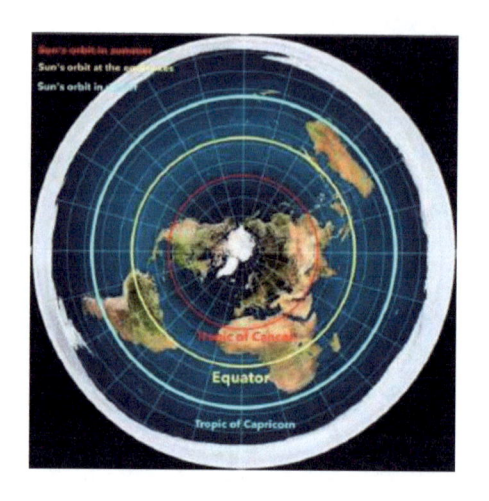

A typical polar azimuthal equidistant globe map from 1892 is shown here.

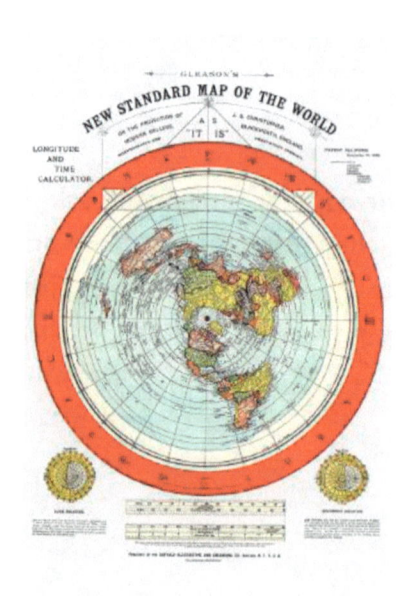

The globe model obviously offers the visitor a quite different situation than the flat earth concept, especially in the so-called "southern hemisphere." Many sailors who traversed the southern seas became lost because of the conflict between the myth of a globe and the reality of a flat world. Based on a globe model, Samuel Rowbotham describes the perils of inaccurate charts as follows:

Navigators to India have frequently thought they were east of the Cape when they were really west, causing them to crash ashore on the African coast, which, by their calculations, was behind them, in the southern hemisphere. The Challenger, a superb frigate, had this catastrophe in 1845.

How did Conqueror, the ship belonging to Her Majesty, become lost? How have so many other noble ships that were absolutely sound, perfectly manned, and expertly navigated, been sunk in calm weather, not just in the dead of night or in a fog, but also in the light of day and in the sun—in the former case on the coasts, in the latter on submerged rocks—from being "out of reckoning," under circumstances that, up until this point, have defied all logical explanations

Rowbotham details the British ship Challenger's round of the southern hemisphere, although indirectly through Antarctica. According to the globular earth model, Antarctica's circumference should be around 14,460 statute miles. The Challenger, on the other hand, traveled for three years and approximately 69,000 kilometers. If the world were a globe, the ship could have circled Antarctica more than four times at that distance. But a flat earth is the only explanation that makes sense for Antarctica's 69,000-mile voyage around the globe.

Rowbotham quotes explorers who discovered that their globular earth-based charts almost always led them astray in the southern seas " Every day, we discovered ourselves between 12 and 16 miles ahead of our reckoning by observation.
An further southern seafarer said: "We discovered ourselves 58 miles to the east of where we had calculated two days earlier based on our views at midday.

The vastness of space between the lines of longitude as one moves further south on a flat globe was not taken into consideration in the spherical earth maps. As a result, the location on the chart that was projected using dead reckoning was discovered to be off by several degrees from the mariners' actual position on the earth, which was established using celestial navigation and a clock. One sailor described how far off course they were in the south seas due to the inaccurate maps, which suggested a spherical earth.

Our chronometers recorded 44 miles more westward motion on February 11th, 1822, at midday in latitude 65.53 S, than the log recorded in three days. On April 22, 1822, in latitude 54.16 South, our longitude, measured by chronometers, was 46.49, and by dead reckoning, it was 47° 11': Our longitude by chronometers at noon on May 2, 1822, in latitude 53.46 S, and by D.R., was 59° 27' and 61° 6', respectively. The 14th of October, in latitude 58.6 and longitude 62° 46' by chronometers and 65° 24' by account. By chronometers, the longitude at latitude 59.7 S. was 63° 28', while by accounts it was 66° 42'. By chronometers, the longitude at latitude 61.49 S. was 61° 53', according to accounts, 66° 38'.

The skippers of the ships in the southern seas could only surmise that the currents were to blame for the differences between their dead reckoning plotted on their maps and their real positions, which were verified by accurate chronometers and sextants. However, Rowbotham argues that currents could not reasonably be the source of the navigational mistakes because they were evident whether the ships were sailing east or west.

According to Lieutenant Wilkes, the commander of the American exploration mission, he traveled 20 miles to the east of his estimated location in latitude 54° 20' S in less than 18 hours. He provides further examples of the same occurrence and, like nearly all other navigators and authors on the subject, he blames currents, the velocity of which in latitude 57° 15' S. was equal to 20 miles per day, for the discrepancies between real observation and theory. With their training and conviction in the rotundity of the globe, the expedition leaders were, of course, unable to think of any other explanation for the discrepancies between chronometer and log data except the presence of currents. The fact that the outcomes are the same whether the path is traveled east or west, however, renders such an argument completely invalid. Since the water in the southern region cannot flow in two opposing directions at once, despite the fact that a number of local and variable currents have been observed, it is impossible to prove that they are to blame for the differences between time and log results that are so frequently seen in high southern latitudes. The conclusion is one that was forced upon us by the totality of the evidence, showing that the degrees of longitude in any given southern latitude are greater than the degrees in any latitude closer to the northern center. This proves the already amply demonstrated fact that the earth is a plane with a northern center, in which the degrees of latitude are concentric and the degrees of longitude are diverging lines that are continuously in motion.

The sun, moon, and their paths are shown on the map below. They would move forward along their course from the Tropic of Cancer for the northern summer (the southern winter) to the Tropic of Capricorn for the northern winter (the southern summer).

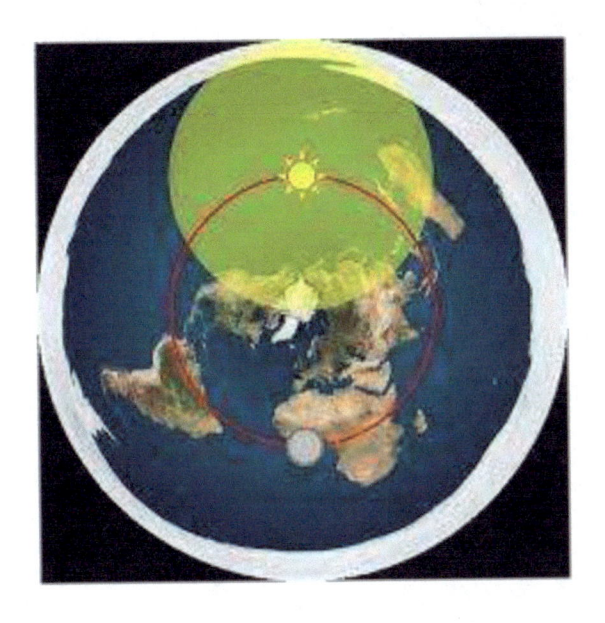

The majority of people should think that the globe is round, according to the world's governments. However, governments turn to a polar azimuthal equidistant flat earth map when they seek to understand how to deal with problems arising from the actual layout of the planet. In fact, the United States Military utilizes hundreds of troops and civilians to create precise maps of deployment zones. These maps hardly ever get public. An image of a polar projection of an azimuthal equidistant map in the White House Situation Room is shown below from an official video tour of the room that the White House produced and made available online. The United States government wants to know that the world is flat when it matters most.

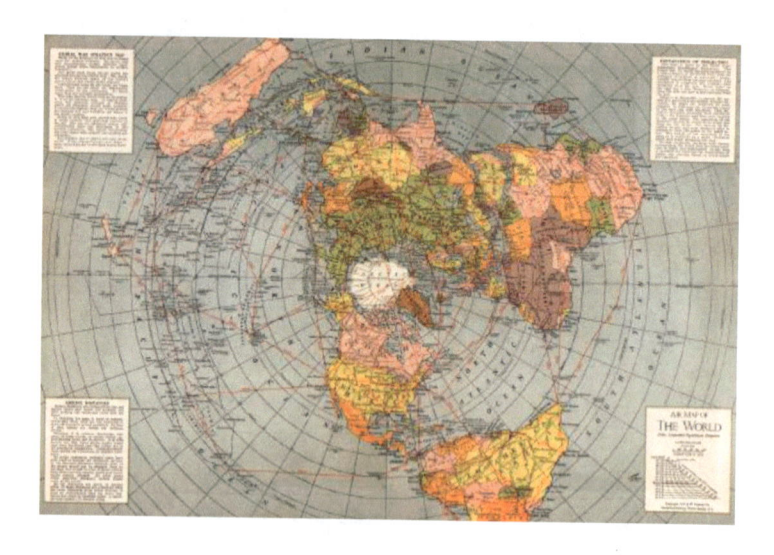

Crushing Gravity

According to the heliocentric hypothesis, the planet is rotating at the equator at a speed of around 1,000 miles per hour. The heliocentric researchers encountered an issue with their hypothesis. How could they explain why neither people nor animals nor objects experience the centrifugal force of the earth's rotation? With the help of his theory of gravity, Isaac Newton rescued the day. According to Newton's theory of gravity, a centripetal force opposes the centrifugal force of the hypothetically rotating earth. A spinning earth requires gravity. However, the earth is not rotating. There is no centrifugal force on a flat earth that is motionless. There can be no centripetal force if there is no centrifugal force. Because a flat, unmoving globe does not require the centripetal force of gravity, there is no such thing as gravity.

The foundation of Newton's theory of gravity is the idea that all objects are drawn to all other objects according to their masses. This "rule" of science, according to Eric Dubay, has never been established and cannot be seen.

There is no place in the natural world where huge things would exhibit the magnetic-like pull that gravity is said to have. There is no instance of a large sphere or other shape in nature where its sheer size forces smaller items to adhere to it or revolve around it! Nothing on Earth is sufficiently large to demonstrate that even a dust-bunny will adhere to or orbit it. You will notice that nothing clings to or circles a spinning wet tennis ball or any other spherical item when you try spinning it with smaller objects on top of it. It is hearsay, not science, to assert the existence of a physical "rule" without providing even one concrete, empirical example.

Dubay is true that gravity is an ancient religious myth and not based on science. People accept it as fact merely because modern-day witch doctors—whom we refer to as "scientists"—say it is. Dubay goes on to challenge gravity's seemingly mystical properties, as it simultaneously acts as an attraction force that causes adhesion and a force of suspension that causes orbit.

Now, even if gravity existed, why would it make people stay to the Earth and force planets to orbit the Sun? Either the Earth should be pulled into a collision with the Sun by gravity, or mankind should float in suspended circular orbits around it! What kind of sorcery is "gravity," if it can hold people's feet to the ball-Earth while forcing the planet to orbit the Sun in ellipses? The same cause is given to both effects, despite the fact that they have quite distinct impacts.

The centrifugal force of the rotating globe is perfectly balanced against the gravitational pull near the equator, according to the heliocentric model. All people and things are said to be perfectly balanced by gravity, which pits their mass against the spinning earth's centrifugal force to keep them linked to it.

The issue with the gravitational theory is that it assumes that all people and objects are attracted to the earth in the same way regardless of where they are on the planet. That implies that the gravitational force at the North Pole and the equator are equivalent. If the earth is spinning as claimed, then it offers a very substantial concern. This is due to the fact that the centrifugal force diminishes with distance toward the north pole, where it eventually zeroes out because the north pole is the axis of the ostensibly rotating globe. On a globe, the circumference parallel to the equator decreases as you move north or south of it. As a result, the globe would spin more slowly away from the equator at those higher northern and southern latitudes than it does near the equator. For instance, the planet should spin at around 700 miles per hour at the 45 degree north latitude. Reduced spin speed is accompanied by a corresponding decrease in centrifugal force. This disproves the myths of the rotating world and the mysterious power of gravity.

Objects do, in fact, weigh differently in the equator and the North Pole, others will point out. The discrepancy, however, is the opposite of what the relationship between centrifugal force and the theory of gravity would predict. At the North Pole, objects are somewhat lighter than they are at the Equator. The explanation is due to the fact that items at the equator weigh a little bit more due to increased air pressure as one gets closer to the equator.

There is no such thing as gravity, and a flat world does not require gravity. Density prevents items from floating off the earth's surface. People and things cannot float off the ground because they are heavier than air. Naturally, there are some gases that are lighter than air and float above the surface. Helium balloons are a common sight in the sky. Helium balloons are lighter than air, therefore everyone is aware that they are not some kind of anti-gravity gadget. Instead, they float in the air. Why do people not understand that apples fall from trees to the ground, not because of gravity, but because apples are denser than air? They believe in the mystical force of gravity, not because it has been proven true, but because they have been brainwashed into believing in it. Gravity does not exist. David Wardlow Scott explains:

Any item that is heavier than air and is not supported by anything else will inevitably start to fall from its own weight. The ripe apple, even Newton's well-known Woolsthorpe apple, releases its grip on the stem and falls to the earth out of necessity since it is heavier than air, completely disregarding any pull of the Earth. Because if there were such an attraction, why doesn't the Earth draw the rising smoke, which is not nearly as heavy as the apple? The solution is straightforward: because the smoke is lighter than the air, it ascends rather than falling. The sooner gravity is consigned to the tomb of all the Capulets, the better for all social strata. Gravitation is only a ruse used by Newton to try to establish that the Earth rotates around the Sun. With a little modification of Byron's famous comments, it is still true that "[m]ortals, like moths, are sometimes drawn by dazzle." He covered his idol in the tawdry glitter of fake science, knowing well well how to deceive the unthinking masses. And while Seraphs may despair, foolishness succeeds. In the ingenious hocus-pocus game of "heads I win, tails you lose," Newton acquired his renown as the public lost their senses.

As a scientific hypothesis, gravity is more like a pagan religious doctrine. The "scientists" who support gravity are like priests in a religious cult who have elevated the gravity inventor to sainthood and adhere to the gravitational theory as if it were some sort of sacred doctrine.

Sir Isaac Newton, a physicist who wrote the influential three-volume Philosophiae Naturalis Principia Mathematica (The Mathematical Principles of Natural Philosophy) in 1687, is credited with founding the widespread, cult-like belief in gravity. Newton presented his laws of motion and the theory of universal gravitation in the book. The majority of scientists believe that Newton was the key theorist who first proposed the concept of gravity.

There is no scientific evidence to back up gravity. So where did the theory come from? According to S. Pancoast, Isaac Newton popularized "gravity" as a word for what the Kabbalah refers to as "the rule of attraction and repulsion." As S. Pancoast puts it:

Pythagoras was never allowed to speak in public about what he knew and believed. Instead, he taught his closest students all the glories of his philosophy while adhering to the strictest confidentiality rules. It was forbidden for

Pythagoras to share this information since doing so would expose the sanctuary's greatest secret, the law of attraction and repulsion. Newton's study of the Kabbalah more than a thousand years later helped him uncover these forces.

Most people are unaware of the fact that Isaac Newton was a religious mystic who was well-versed in Judaism and Jewish writings. According to Aron Heller of the Times of Israel, Newton "studied Jewish philosophy, the mysticism of Kabbalah, and the Talmud. He also learned how to read Hebrew and scrolled through the Bible. In fact, according to a writer, "7,500 pages of his [Newton's] theological ideas, written in his own hand, are digitalized at Israel's national library at Hebrew University.

Before they were transferred to the National Library of Israel at Hebrew University in Jerusalem, we can't seem to find out who was in charge of looking after Isaac Newton's religious works. Abraham Yahuda, a Zionist Jew, was the one who acquired and preserved Newton's theological papers. Yahuda searched the globe for Isaac Newton's religious works. Why was Yahuda so intrigued by Isaac Newton's writings? Yahuda "began to try to buy the Newton documents and wrote to [his wife] Ethel on July 28, 'I am excited with the notion of having them,'" according to Sarah Dry. He wrote extensively about the Bible and Jews, the Cabbala, and many Jewish issues.

Albert Einstein's contemporaries Yahuda and Einstein discussed Newton's esoteric theological works. Einstein was a Zionist Jew as well. Given their shared belief in the validity of mysticism, Einstein and Newton shared many religious beliefs. Sarah Dry gives the following intriguing detail concerning Einstein's perspective on Isaac Newton's release of his theological writings:

Just two weeks before he passed away in 1955, Einstein chatted with the scientific historian I. B. Cohen about a variety of topics, including Newton's religious works. According to Einstein, the fact that Newton had "shut them all up in a box" was crucial since it showed that he was aware of their flaws. It was "clear" that Newton did not want these conjectures to be made public while he was still alive. With "some emotion," Einstein expressed his desire for them not to be made public at this time.

Why did Einstein feel so strongly that Newton's religious works should not be made public? Because they would abandon the game, those texts. Isaac Newton's theological works would reveal to the public the Jewish Kabbalah as the religious foundation for his theory of gravity. Because of this, Zionist Abraham Yahuda searched the world to gather Newton's theological papers, and they are now safely housed in Jerusalem in the Hebrew University's National Library of Israel. Dry goes on to say:

Like Stokes and Adams before him, Einstein studied Newton's personal papers in an effort to learn as much as he could about his research process, which he refers to as "the formative evolution" of his physics work. Newton's development of his physics and religion are implicitly connected by Einstein, who suggests that by studying one, we could learn more about the other.

Theological works of Newton were seen by Einstein as "the formative evolution of his work in physics." Regarding Isaac Newton's theology, Einstein said the following in a letter to Yahuda:

I find Newton's works on biblical issues to be extremely fascinating since they offer in-depth insight into the intellectual characteristics and techniques of this significant guy. Newton is quite confident that the Bible was inspired by God, a view that is oddly at odds with his critical cynicism of religious institutions. This assurance gives rise to the steadfast belief that the seemingly enigmatic passages of the Bible must contain significant insights, which may be made clear by understanding its symbolic language. Newton uses his acute, methodical reasoning, which is based on the meticulous use of all the resources at his disposal, to seek this decipherment or interpretation. We do have in this area of Newton's works on the Bible drafts and their repeated modification; these mostly unpublished writings therefore allow a highly interesting insight into the mental workshop of this unique thinker. While the early development of Newton's lasting physics works must remain shrouded in darkness because Newton apparently destroyed his preparatory works, we do have in this domain of Newton's works on the Bible drafts and their repeated modification.

The true source of Sir Isaac Newton's idea of gravity is his study of the Kabbalah. According to Marshall Hall, "[w]hile Kabbalism is almost generally linked with Judaism, there are so-called "Christian Kabbalists," including such like Newton, Dee, Kepler, Shakespeare, Cardinal Nicolas of Cusa, and a lengthy list of Theosophists, Rosecrucians, Masons, Crowleyites, etc. All of these and more have promoted or do so now the mysticism of the Kabbalah. With his obscure "mathematical" conceptions, Newton, according to Hall, "in short, ensured the acceptance of the Copernican Model for two centuries; concepts upon which others—Einstein through Sagan, et al—could built today's Pharisee Cosmology."

Gershom Scholem, who served as the first Professor of Jewish Mysticism at the Hebrew University of Jerusalem, is regarded as the father of the contemporary, academic study of Kabbalah. According to Scholem, Jewish mystics have known about gravity since ancient times. Newton would have undoubtedly encountered this when researching the Jewish Kabbalah. Gravity was linked by Gershom Scholem to the last Heh, the Shekhinah, the Presence, the alchemical Earth, and the divine daughter. Take note of the intriguing relationship between our planet's gravity and the alchemical Earth!

The Kabbalah & Magic of Angels by Migene Gonzalez-Wippler discusses the fundamental link between the Jewish Kabbalah's mysticism and the theory of gravitation.

Cause and effect and the law of momentum come together to form gravity. It drives the expansion of the universe and collaborates with dark matter to preserve the equilibrium of galaxies, planets, and stars. It is compared to Tiphareth [also known as Tifereth], the sixth sphere of the Tree of Life, which is the center of the Tree underneath Kether, the first sphere, for this reason. The sun, which is at the core of the solar system and uses gravity to keep the planets in stable orbits, is often referred to as tiphareth.

Gonzalez-Wippler alludes to a pictorial representation of the Jewish deity Ein Sof called the Tree of Life. The sefirot are represented on the Tree of Life as spheres, which is no accident. The "scientific" notion of gravity has its roots in the Tree of Life (Ein Sof). If there is a force, it must exist to explain why people and objects do not fly off into space, in order for the myth of the spherical earth to be accepted.

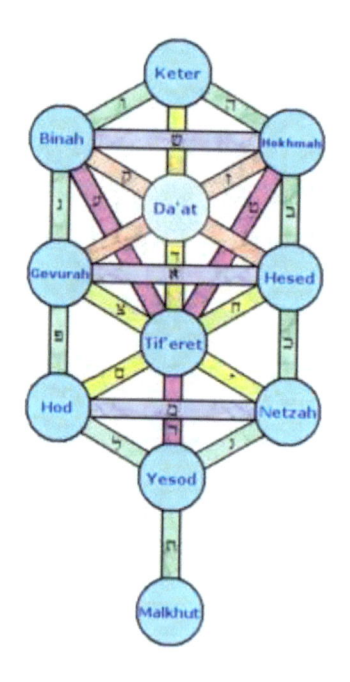

The Tree Of Life (Ein Sof) From the Jewish Kabbalah

Gravity does not need to be demonstrated or even to make sense as long as it is supported by a front of professionals referred to as "scientists." However, there is nothing scientific about gravity, make no mistake about it. It derives from the Jewish Kabbalah and is pure hedonism. No disagreement with their tenet of gravity is tolerated by the priest-scientists of the heliocentric religion. Gravity "is associated with Tiphareth [a/k/a Tifereth], the sixth realm of the Tree of Life," claims Migene Gonzalez-Wippler. According to the explanation provided below, this indicates that gravity is not just a quality of the Jewish god Ein Sof but also a separate Kabbalistic deity.

Ein Sof, the Kabbalah's deity, has 10 different characteristics (sefirot). Each sefirah (plural of sefirot) is a characteristic of Ein Sof, as well as an anthropomorphic aspect of that one god. Aside from that, every sefirah has its own deity or goddess. Three triads made up of three sefirot each and depicting the three main anthropomorphic elements of the mystical body of Ein Sof are formed from the first nine sefirot (plural of sefirah). The Shekinah, also known as Malkuth, is the eleventh sefirah and is not one of the three triads.

The three sefirot that make up the lowest third of the Kabbalah's pagan god Ein Sof are Netzach (endurance/victory), Hod (majesty/glory), and Yesod (Foundation). Ein Sof's right and left legs are Netzach and Hod, while his phallus is Yesod. According to the Kabbalah, the Shekinah (also known as Malkuth), the last Sefirah, receives the light and strength of the Sefirot through the phallic deity Yesod. The Kabbalah's overtly sensual portrayal of the Jewish god includes this phallic deity. In The Encyclopedia of Jewish Myth, Magic, and Mysticism, Rabbi Geoffrey W. Dennis writes: "The Zohar provides many interpretations based on the idea of God's genitalia.

The Jews embraced the phallic religion of Judaism while they were being held captive in Babylon. The Jewish liturgy's emergence of esoteric sexual implications is explained by Dan Cohn-Sherbok and Lavinia Cohn-Sherbok as follows:

The ninth Sefirah, Yesod, from which all the higher Sefirot poured into the Shekinah as the life power of the cosmos, was also the subject of hypotheses using phallic symbolism. The action in prayer that was portrayed as copulation with the Shekhinah was described in subsequent Hasidic literature using sensual vocabulary.

Shuckling, also known as shokeling or shoklen, is the motion reported by Dan and Lavinia Cohn-Sherbok as occurring during prayer. Additionally, it is known as davening, which is just Yiddish for praying.
Jews mimic copulation in a sexual union with the Shekhinah via shuckling. Shuckling is a sign of Judaism's phallic religion, which features liturgical customs and prayers with hidden sexual connotations. Regarded as the father of Hasidic Judaism and a leading figure in Jewish theology, Baal Shem Tov. "Prayer is coupling with the Shechinah," according to Baal Shem Tov. Swaying back and forth is a sign of mating during prayer. Explained by Rabbi Eli Malon: By "prayer," he [Baal Shem Tov] meant the physical rocking back and forth that is typical of traditional Jewish prayer, which is evocative of sexual activity. Jews see the swaying as a symbol of copulation with the goddess Shekinah. Nathaniel Kapner, a former Jew also known as Brother Nathaniel, affirms the secret significance of the Jews' swaying:

Pay special attention to how the rabbis thrust their pelvises and penises back and forth during a prescribed prayer movement known as "davening," in which the Jew mates with the "Shekinah" to produce an amorous union with the "Ein Soph," the Kabbalistic male emanation of their false deity.

Jews are commonly aware of the esoteric sexual connotation of shuckling, while Gentiles are not informed of this meaning. In a piece written specifically for a Jewish audience and published in the San Diego Jewish Herald in 2013, Rabbi Michael Leo Samuel publicly described the shuckling of Hasidic Jews in front of a Victoria's Secret lingerie store. He questioned if the davening of the Hasidic Jews required a visual help. Baal Shem Tov's writings on the significance of the swaying during Jewish prayer were summarized by Samuel:

With the Shechinah, prayer is zivug (coupling). Similar to how there is movement at the beginning of a marriage, there must be movement (swaying) at the beginning of prayer. After that, one can remain motionless and firmly cling to the Shechinah with great deveikut (faith in God). However, by swaying, one can experience severe punishment. The reason being that you ask yourself, "Why do I move myself? Maybe it's because I can definitely see the Shechinah standing before me. Your level of hitlahavut (enthusiasm; rapture) will increase as a result. (Original parentheses in parentheses)

Rabbi Samuel appeared to be angry that Jews were shushing in front of a Victoria's Secret lingerie store because it communicated too much to the outside world about the esoteric significance of the Jewish religion. They genuinely believe that no one is paying attention, according to Rabbi Samuel, when Hassidic Jews pray in front of Victoria's Secret. They act in a manner similar to a little child who covers his ears while screaming, believing that no one else can hear him.

Sex magic is a deep undercurrent of phallic worship introduced by the Kabbalah into traditional Judaism. A branch of Judaism's hidden teaching is the sex magic. The idea that the mystic might achieve atonement by a "heroic" resolve to perform evil is a prevalent one seen in secret organizations. According to a hidden rabbinic theory, committing evil has spiritual benefits that may be attained by accepting them. That clarifies what Jesus meant when He told the Jews, "You are of your father the devil, and you will fulfill the lusts of your father." John 8:44. There is widespread pederasty among Jewish clerics, as there is with the clergy of other phallic religions. In her book Solving the Mystery of Babylon the Great, the author provides documentation of rabbinic pederasty.

The notion of the Zaddiq [Jewish mystic or saint], which is better known by the Hebrew term Yeridah zorekh Aliyah, essentially the descent for the purpose of the ascension, the transgression for the cause of r epentance, is described by Moshe Idel in Hasidism Between Ecstasy and Magic. Due to its fundamental similarities to Zoharic and Lurianic Kabbalah, this model has received a lot of attention. This model was crucial to Hasidic philosophy. The fundamental tenet of the "black magic" belief system is that idea. Babylon is the origin of this occult "black magic" ideology. The books Sifreihlyyun, Sefer ha-Bahair, and the Hikoth Yesirah, often known as the Sefer Yetzirah, are the earliest sources for this Babylonian black magic in Judaism.

The Shekinah and the Tif'eret are united by the Yesod (Jesod). The child of Hokhmah and Binah is Tif'eret. Two of the three sefirot of the divine head of the mystical body of the Ein Sof are the Hokmah and the Binah (Kether is the third sefirah). In addition to being a deity in and of himself, Tif'eret also symbolizes the torso and heart of the Kabbalah god Ein Sof.

The "tree of life" picture provides a visual portrayal of the relationships between the various sefirot within the Ein Sof and shows the 10 sefirot as they are given in the Sefer Yetzirah (Book of Formation). Be aware that Shekinah and Malkuth belong to the same sefirah. The lone figure in the diagram is that of Malkuth. Also take note of the little spelling variations across sources. In the figure, Chokmah is equivalent to Hokmah, while Jesod is equivalent to Yesod. Jewish academics are quick to point out the many similarities between the gods of Buddhism, Hinduism, and so-called Gnosticism and the gods of the Cabal.

That is hardly unexpected given that they all originate from the same enchanted waters of Babylon. The sixth realm of the Kabbalistic Tree of Life, Tiphareth [also known as Tifereth], is compared to gravity in the Kabbalah. Sir Isaac Newton only provided the world with this pagan religious ideology disguised in mathematical formulae as "science."

Simple Evidence Against Heliocentrism

The idea that the sun may be much smaller than the earth and restricted to the region immediately above its plane is a novel one for the majority of people. Let's explore the accepted theory of the heliocentric model before going through the scientific evidence for it. The earth revolves around the sun once every 365.25 days according to the heliocentric hypothesis. Since there are only 365 days in a year, we get a leap year every four years (4 .25 = 1 day), which adds a 29th day to February to make up for the one day that is lost. The earth is scheduled to complete a full 360-degree circle around the sun in those 365.25 days. At the same time the earth is orbiting the sun, it is completing one 360° rotation every 24 hours.

According to the heliocentric model, the earth will move a little less than one degree in its orbit every day as it revolves around the sun. In other words, the earth will rotate 180 degrees in its orbit around the sun every six months, putting it in opposition to the sun from where it was six months prior. With this model, there is a flaw. Let's imagine that on September 22 at noon in New York, we start watching the sun. The length of a solar day is exactly 24 hours.

The heliocentric model predicts that the earth should complete a precise 360-degree rotation on its axis and arrive at the same location once every 24 hours. At the same time, the earth has rotated slightly less than 360 degrees around the sun each day. Accordingly, on March 21 at the midst of the dark night, 12:00 noon in New York will come after six months.

To recall, on the day of our start on September 22nd, 12:00 noon comes in the middle of the day in New York. The earth must be on the other side of the sun six months later according to the heliocentric hypothesis. Therefore, at noon on March 21, New York will be looking away from the sun on the dark side of the planet. So on March 21 at 12:00 noon, it should be the middle of the night in New York. This phenomenon ought to occur continuously throughout the year according to the heliocentric hypothesis. The heliocentric paradigm is incorrect since we are aware that this does not actually occur.

Our 365-day calendar dictates that we require a leap year once every four years to make up for a lost day. As a result, the earth should have missed 1/8th of its orbit around the sun after six months. The heliocentric model, however, predicts that 12:00 noon in New York will travel a little less than one degree every day until it reaches the midst of the dark night opposite the sun after six months. This is due to the fact that each solar day is exactly 24 hours long, and according to the heliocentric model, in those 24 hours the earth must spin 360 degrees on its axis.

The top priests of heliocentricity have identified the flaw in their paradigm and made modifications to it to account for it. How did they act? They calculated that the earth rotates 360.986 times each day by adding .986 to its 360-degree axis. How did they obtain the .986? They just made it up by multiplying the 360-degree orbit by the 365.25-day year to get .986-degrees. Their issue appears to be resolved with this small mathematical change. They determined that it takes the earth 23 hours, 56 minutes, 4.1 seconds to revolve 360 degrees since they assert that it rotates 360.986 times in a day. It takes the earth 23 hours, 56 minutes, and 4.1 seconds to revolve 360 degrees. A sidereal day is the name given to this shortened day. The duration of time between the positions of the stars in the night sky is believed to serve as the basis for a sidereal day. The length of a sidereal day has no bearing on how the sun moves.

Below is a diagram from Cornell University Department of Astronomy illustrating the contrived 360.986 rotation each day.

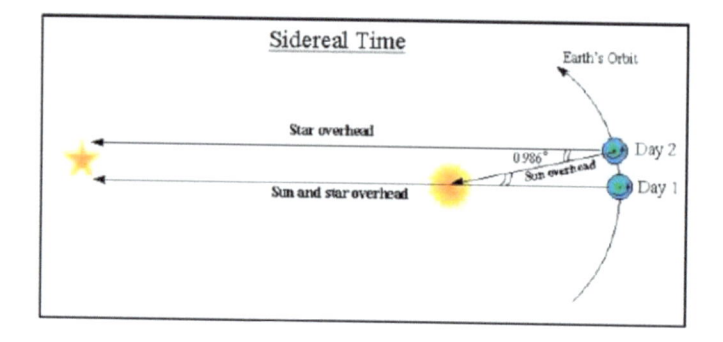

Sidereal Time

Diagram showing Earth's Orbit with Day 1 and Day 2 positions, Star overhead, Sun and star overhead, and a 0.986° angle for Sun overhead.

The first issue with the 360.986 answer is that it deviates from the heliocentric model's accepted wisdom, which asserts that the planet rotates precisely 360 degrees every 24-hour solar day. The traditional heliocentric model calls for a 360-degree rotation every 24 hours since that is how the sun really orbits the earth. Every 24 hours, the sun travels a full 360 degrees around the planet.

The Colorado State University-affiliated teacher resource center for the Annenberg Foundation offers a representation of the globe-spinning planet and notes that "[e]ach day, the world revolves once on its axis, which equals 360 degrees. The globe rotates 360 degrees every 24 hours, according to the online self-study manual from the University of California, Berkeley: "Full turn = 360 degrees. The world spins 360 degrees "once every 24 hours," according to the Department of Astronomy and Astrophysics at the University of Chicago.

According to Professor of Astronomy Courtney Seligman, who has been teaching college astronomy for 39 years, the Earth rotates at a speed of 4 minutes per degree. This is determined by dividing the length of a day, which is 24 hours or 1440 minutes, by the 360 degrees it travels through during one rotation.

Using a figure, Professor Seligman depicted how the earth orbits the sun. He explained the graphic in the manner shown below:

The planet's positions at four different times, each spaced a third of a rotation period apart from the other, are shown by the four blue dots on the right side of the picture below. The numbers to the right of each dot represent the number of spins the planet has completed. The enormous yellow dot on the far left indicates the position of the Sun, and the white dot illustrates how a particular location on the planet changes as it spins to the east (in this figure, counter-clockwise). To make it simpler to understand what is going on, the diameters of the Sun and planet have been inflated, as well as the angle the planet goes through during one spin. a planet's motion throughout a single spin. The Sun seems to move across the sky as a result of the planet's orbiting around the Sun. The Sun seems to rotate the planet by one degree for every degree that the planet goes around the Sun.

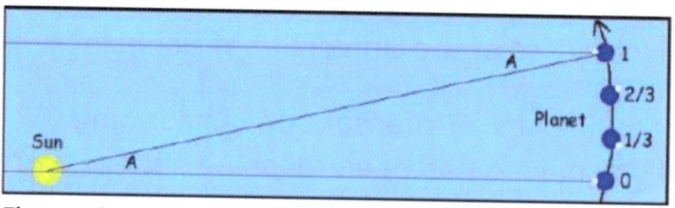

Figure: Diagram by Professor Seligman explaining the supposed 360° rotation of the earth as it orbits the sun

Professor Seligman uses the motion of the earth around the sun as it spins on its axis 360 degrees throughout each 24-hour day to explain the heliocentricity standard model. His illustration backs up the claim that the heliocentric construct shown in the chapter's opening illustration is the norm. The illustration of the earth's heliocentric orbit at the beginning of this chapter came from the John A. Dutton e-Education Institute at Penn State University, and it represents the current de facto accepted heliocentric model. This author annotated the diagram, indicating in the captions where 12:00 noon in New York would fall for each quarter of the earth's annual orbit. The diagram looks to be exactly how Penn State drew it, save from the comments.

39 years of experience teaching collegiate astronomy have given Professor Seligman a thorough knowledge of heliocentrism. His illustration shows how the heliocentric concept is incompatible with reality. As shown in the figure at the beginning of this chapter, if Professor Seligman's design were stretched to include a 180-degree journey of the earth around the sun, we would discover that 12:00 noon would arrive in the middle of the night after six months. We are aware that New York's day officially begins at 12:00 noon every day. In all of New York's recorded history, there has never been a day when noonfall falls in the dead of night. That one fact alone disproves the heliocentric paradigm.

The earth does not orbit the sun, according to Professor Seligman and the vast majority of astronomers. However, they are right when they say that there are 360 degrees and 24 hours between the sun and the earth. Every 24 hours, the sun really completes a 360-degree cycle around the flat planet.

The theory put out by other astronomers that the earth rotates 360.986° every 24 hours is demonstrably false. The sun can be observed to circle the earth once every 24 hours, which disproves this theory. That fact is the foundation of all celestial navigation. Based on the verifiable fact that the sun travels 15 an hour (360 x 24 hours = 15), celestial navigation is based on this. By multiplying 60 minutes by 15 degrees, or 4 minutes per degree, one may convert an hour into minutes per degree.

Sammons, James I., discusses the fundamental ideas of celestial navigation. Like most others, he has been taught to think that the sun moves in its course because of the earth's rotation. But when he says that the sun and the earth have a connection of 360 degrees in a day, he is 100% accurate. In fact, that serves as the fundamental compass for all celestial navigation.

Finding longitude is not as difficult as determining latitude if we have a grasp of three fundamental concepts. The first of these concepts is the connection between time and Earth's rotation. The Earth rotates 360 degrees in an average of 24 hours, allowing a place on its surface to pass from beneath the Sun and then simply return to its initial location. The Earth will complete one full rotation in 12 hours. In six hours, one-fourth. The Earth rotates 15 degrees each hour, or one full circle's worth of degrees, divided by the number of hours in a day. 15° each hour Equals 360° x 24 hours. This may be expanded upon to mean that the Earth rotates one degree in every four minutes. 15° x 60 minutes each hour equals 4 minutes per degree.

For hundreds of years, navigators have used the stars to mark a precise path. It is based on the sun's 24-hour, 360-degree rotation of the planet. Modern seafarers have undoubtedly been trained to think that the spherical globe is rotating. However, when mariners compared their dead reckoning using their charts with their more precise celestial navigation, they frequently discovered that their charts, which were based on a spherical earth, were erroneous (especially in the southern latitudes). The fact that celestial navigation relies on the sun's 360-degree motion in a 24-hour period and that it is functional undermines the validity of the sidereal day concept, which assumes a 360.986-degree rotation of the planet in a 24-hour period.

According to the dictionary, sidereal refers to or is stated in reference to stars or constellations. A sidereal day is a unit of measurement for a star's motion. Applying calculations for the movement of the stars to the movement of the sun is blatantly dishonest. The sun has never navigated the heavens by traversing 360.986 miles in a day. That demonstrates that the added .986 is incorrect. It is a scheme created by the priests of heliocentrism. It is a myth that was created to justify the heliocentric paradigm, which is inherently implausible. The scientists' assertions are the only reason the added .986 has any validity. The observed fact that the sun makes a perfect 360-degree cycle around the globe once every 24 hours casts doubt on their 360.986-degree concoction.

The heliocentric model would have to move high noon every day in order for it to eventually arrive at midnight once every six months, according to that observable fact. The heliocentric hypothesis is false because such an event does not occur. One of the most well-known physicists of the previous century was Max Planck. In 1918, he was awarded the Physics Nobel Prize. He exposed the scientific community's cult-like belief structure. Anyone who has engaged in serious scientific study of any type "understands that over the entrance to the gates of the temple of science are inscribed the words: Ye must have faith," he said. The scientist cannot live without this trait. Today's so-called "scientists" are more comparable to witch doctors who have charmed the superstitious tribe into accepting their gibberish.

Objects in the Distance Can Be Seen Over Water

There is just no way to explain how lighthouses may be seen from such a far distance at sea if, as is believed, the world is a sphere. Modern seafarers can currently encounter this occurrence. Lighthouses and other buildings may be seen from a great distance out at sea, and the only possible explanation for this occurrence is that the planet is flat.

Rowbotham cited several instances of lighthouses that were actually visible from a long distance. If the world were round, the lights would have been well below the level curve and out of sight.

The Egerö Light, on the west tip of Island, on Norway's south coast, is equipped with the first order of dioptric lights, has a visibility range of 28 statute miles, and is 154 feet above sea level. The correct calculation will reveal that this light has to be 230 feet below the horizon. The Dunkerque Light, located on the southern French coast, stands 194 feet tall and can be seen from 28 statute miles away. According to the standard computation, it should be 190 feet below the horizon.

The Cordonan Light is located on the River Gironde on the west coast of France. It is visible from 31 statute miles away, and at a height of 207 feet, its dip below the horizon is estimated to be around 280 feet. At a distance of 28 statute miles, the Madras Light on the Esplanade, which is 132 feet high, should be more than 250 feet below the horizon.

The Port Nicholson Light in New Zealand, which was built in 1859 and is located 420 feet above high water, has a 35 statute mile visibility range. The water should be 220 feet below the horizon if it is convex.

In Newfoundland's Cape Bonavista, the light is 150 feet above high tide and may be seen for 35 statute miles. These numbers indicate that the earth should be buried 491 feet below the sea horizon when accounting for the planet's rotundity.

Rowbotham demonstrated that the planet is a flat plane using common, easily accessible sources. He cites several cases from Alexander G. Findlay's Lighthouses of the World, the leading work on lighthouses at the time, as an example (1861). Findlay was a well-known geographer and hydrographer who published a number of reference works that were very helpful to seafarers all around the world. Findlay was an expert in hydrography without a doubt. His reference book Lighthouses of the World was a trusted resource for sailors. Since the seafarers' very lives depended on the correctness of the information in that text, it was crucial that the entries in that reference book be exact. Although the content was undoubtedly correct, Rowbotham showed that if the planet were a globe, several of the lights listed in that paragraph shouldn't have been seen from the distances given. The flat earth hypothesis is the only one that makes sense. The Bidston Lighthouse is one illustration that Rowbotham provides from Lighthouses of the World, page 39.

The Bidston Hill Lighthouse, close to Liverpool, is 228 feet above high water, has one brilliant fixed light, and is visible for 23 nautical miles, or very nearly 27 statute miles, according to the same reference, at page 39. The downward curvature is 352 feet after subtracting the observer's height of 4 miles (10 feet above the water), squaring the remaining 23 miles, and multiplying the product by 8 inches. From this, subtract the light's altitude of 228 feet, leaving 124 feet as the distance at which the light should be below the horizon!

Bidston Lighthouse.

An illustration of the Bidston Lighthouse taken from Sailing Directions from Point Lynas to Liverpool is shown above. In 1873, a new lighthouse that was constructed close by took the place of the Bidston Lighthouse. The page from the book Lighthouses of the World that Rowbotham quoted is reproduced below.

Please take notice that Findlay defined the least distance at which a lighthouse may be seen in clear weather from a height of 10 feet above sea level as the distance at which the light can be seen in his book. Given that Findlay specifies the minimum distance in clear weather, it follows that in clear weather, the lighthouse may be visible from a distance greater than that specified. Additionally, the distances mentioned are based on clear skies. Of course, the distance from which a lighthouse could be seen would be reduced if the weather was cloudy owing to humidity or other factors.

Rowbotham gave the example of how ship passengers in St. George's Channel could view the lights on both shores simultaneously from the center of the channel. Each lighthouse was 30 miles from the ship traveling through the center of the canal due to the lighthouses' 60-mile separation from one another. The absence of such a phenomena would be guaranteed if the world were a globe. Rowbotham clarifies:
There are at least 60 statute miles between Holyhead and Kingstown Harbour, which is close to Dublin, across St. George's Channel. Passengers frequently observe the Light on Holyhead Pier and the Poolbeg Light in Dublin Bay when in and for a substantial distance beyond the Channel's center [as illustrated in the diagram below].

But the lighthouses would not have been seen at all if the planet were a globe. In fact, even after accounting for the observer's height above the sea at a distance of 30 miles apart from the ship, each lighthouse would be more than 300 feet below the horizon due to the curvature of the hypothetical spherical earth.

Rowbotham clarifies:

A ship in the center of the Channel would be 30 miles from each light, and if the observer were on deck and 24 feet above the water, the horizon on a globe would be 6 miles away. The Holyhead Pier Lighthouse emits a red light at a height of 44 feet above high water, and the Poolbeg Lighthouse displays two bright lights at a height of 68 feet. The distance from the horizon to Holyhead on the one hand and to Dublin Bay on the other would be 24 miles after subtracting 6 miles from 30. A declination of 384 feet is indicated by multiplying the square of 24 by 8 inches. The red light on Holyhead Pier is 44 feet high, whereas the lights on Poolbeg Lighthouse are 68 feet in the air. As a result, according to the accompanying diagram, if the earth were a globe, the former would always be 316 feet and the latter 340 feet below the horizon. The line of sight H, S, would be a tangent that touched the horizon at H and went above the top of each lighthouse by more than 300 feet.

Rowbotham's conclusions were challenged by evolutionist and world-is-a-sphere believer Robert Schadewald. In citing Rowbotham's book Zetetic Astronomy in support of his estimates for the Ryde Pier Lighthouse, Schadewald wrote:

This conclusion [that the Earth is flat] is strongly supported by seafarers' observations of certain lighthouses. The current theory of the Earth's rotundity would make this completely impossible in situations when the light is fixed and highly dazzling. For instance, the Ryde Pier Light, built in 1852, is characterized as a strong fixed light on page 35 of "Lighthouses of the World" as being visible from an altitude of 10 feet at a distance of 12 nautical or 14 statute miles. The horizon would be 4 statute miles away from the observer at an altitude of 10 feet. The square of the next 10 statute miles will result in a 66-foot fall or curve from the horizon. The amount that the light should be below the horizon is 45 feet when the altitude of the light, which is 21 feet, is subtracted.

What interpretation did Schadewald, a skeptic, make of Rowbotham's findings? Rowbotham's computations were entirely accurate, according to his examination of them. Shadewald attempted to explain the visibility of the Ryde Pier Lighthouse by introducing light refraction. His theory of refraction, however, was unable to account for the apparent sight of the lighthouse, which, according to a spherical earth, should have been below the horizon. Schadewald was left with a puzzled expression.

Despite not accounting for atmospheric refraction, Rowbotham's calculation is valid. However, even a generous adjustment of 1/7 of the dip for refraction does not resolve the sphericity issue.

Rowbotham's conclusions about the aforementioned Bidston Hill Lighthouse were closely examined by Schadewald. Schadewald was once more forced to concede that Rowbotham was right. The visibility of the lighthouses was not accounted for by Schadewald's effort to introduce atmospheric refraction into the explanation of Rowbotham's findings.

Once more, Rowbotham's math is accurate, but the issue with sphericity cannot be resolved even with a liberal adjustment for atmospheric refraction. In Zetetic Astronomy, Rowbotham provided roughly 20 of these instances. He asserted that "many further cases may be presented from the same work, shewing that the practical observations of seafarers, engineers, and surveyors utterly reject the idea that the world is a globe." Being the skeptical reader that you are, you undoubtedly question Rowbotham's utilization of Lighthouses of the World. Bresher also pondered. But after consulting the material, he saw that Rowbotham had accurately quoted the published data.

Even if the lighthouses' visibility demonstrated that the world was flat, Schadewald did not agree with Rowbotham's conclusion. Schadewald was left contending that Rowbotham selectively chose visible lighthouses from Findlay's Lighthouses of the World, while in reality they should have been invisible and below the horizon. He reasoned that both a spherical and a flat world could show the majority of the lighthouses. Therefore, he labeled those lighthouses as "anomalies" that, on a spherical world, should have been out of sight, below the horizon. As said by Schadewald:

This tactic is a prevalent trait of persons who are motivated to persuade others of their viewpoint through whatever means necessary. As a result, many creationist evangelists search the scientific literature for evidence that doesn't seem to support the mainstream theory. The public is then shown these abnormalities as representative, just like Rowbotham did with his peculiar lighthouses. Of fact, looking for lighthouses is simpler than trying to build a creation model. Contrary to claims to the contrary, the biological universe has no predictive "creation model."

Rowbotham is criticized by Schadewald for citing lighthouses as evidence of a flat earth when they should be invisible on a spherical globe, claiming that he could only provide 20 examples.
Schadewald is off-target. As asserted by Rowbotham, those 20 lighthouses existed, and Schadewald did not contest that they were visible. How does he respond to those realities? He refuses. He shrugs his shoulders and says, "Beats me." He goes on to speculate that "maybe" the lighthouses were spotted under exceptional circumstances or that "possibly" further observations might not support the previously reported findings. That is speculation rather than science. Shadewald is aware that none of those statements is plausible because many people attempted to disprove Rowbotham's assertions at the time his book was published. His adversaries might have quickly determined if the lights were or were not visible from the specified distances. Even one of these individuals, M. R. Bresher, is

mentioned by Schadewald as having to acknowledge the veracity of Rowbotham's lighthouse data. In Schadewald's conclusion, a guy who has accidentally discovered the truth appears to stand up, dust himself off, and carry on with his journey without giving it much consideration.

What about the peculiar lighthouses built by Rowbotham? amazes me. Perhaps the observations that were recorded were taken in an uncommon setting. Perhaps additional observations would not support the stated anomalies for those lighthouses that are still in service. However, several of Rowbotham's lighthouses are likely to have been abandoned, demolished, or weathered by this point. We will never be aware of these. Nobody will ever discover light if they just look for strange lighthouses, that much is clear.

Despite the fact that many lighthouses have been abandoned, many more continue to stand as mute witnesses to the flatness of the world. As an illustration, the Bell Rock Lighthouse is still operational today. The oldest lighthouse in the British Isles, it is a feat of engineering that has rescued numerous ships and lives. Off Scotland's east coast, the lighthouse is situated on a reef that is buried in the water. It is about 10 miles from shore. The BBC named it as one of the seven marvels of the industrial age because to how challenging it was to build a lighthouse in such a violent maritime environment. In 1811, it was first illuminated. It is the world's oldest sea-washed lighthouse still standing. The Bell Rock Lighthouse's light is 90 feet above high tide, and ships may see it from the sea for a distance of 14 nautical miles, according to Alexander Findlay's authoritative reference work Lighthouses of the World (which is a little more than 16 statute miles). Remember that Findlay specified "the least distance" at which the lighthouse might be seen in good visibility from a height of 10 feet above the ocean.

A seafaring mariner would not be able to see the light from 16 statute miles away if the planet were a globe. If the earth were a globe, the light at the top of the lighthouse would be about 6 feet below the horizon, according to calculations that involve squaring the miles, multiplying by 8 inches to account for the curvature of the earth, and accounting for the fact that a mariner would be about 10 feet above the water. But we do know that seafarers may see the light in good visibility from a "minimum" of 16 statute miles. The Bell Rock lighthouse's light may be seen from a ship's deck 16 statute miles away only on a level earth. On a spherical earth, it is impossible to see the Bell Rock Lighthouse from the minimal viewing distance in good weather.

Figure: Bell Rock Lighthouse

The Fourteen Foot Bank Lighthouse is another illustration of a working lighthouse that serves as a sentinel demonstrating that the world is flat. In the middle of Delaware Bay, 12 miles from Bowers Beach, that lighthouse was constructed. At 59 feet above sea level, the lighthouse is visible from ships at sea 14 statute miles away (12 nautical miles).

 A mariner aboard a ship at sea would not be able to see the light from a distance of 14 statute miles if the planet were a globe. If the earth were a globe, the light would be 8 feet below the horizon, according to a calculation that squares the statute miles, multiplies by 8 inches for the planet's curvature, and accounts for the fact that a mariner would be about 10 feet above the ocean. The lighthouse may be seen by seafarers from a distance of 14 statute miles, as far as we are aware. The focal plane of the light is 59 feet away, according to the New Jersey Lighthouse Society. The lighthouse is still in use and is identified on current nautical charts as "GP Fl (2) 20 sec. 50 ft.

[sic] 12 m Horn," which informs mariners that the station is equipped with a fog horn and that the light is a group flashing light (2), every 20 seconds on a 59 foot tower, visible 12 nautical miles [14 statute miles] at sea. A mariner on a ship's deck could only view the Fourteen Foot Bank Lighthouse from 14 statute miles distant if the world were flat. There is enough proof that the planet is flat. Rowbotham's tests may be very readily reproduced in modern times. For As an illustration, the Chicago skyline may be seen in the image below. seen plainly. Joshua Nowicki captured the image as he stood. near Grand Mere Park, Michigan, about 57 miles away. Chicago is far away, across Lake Michigan.

Figure : Photograph of Chicago taken by Joshua Nowicki, as he stood at Grand Mere Park, Michigan, 57 miles away.

The city of Chicago would be below the horizon if the world were a globe. Only if the globe is flat could Chicago be seen from the western shore of Michigan. A map of the 57-mile route from Grand Mere Park, Michigan, to Chicago, Illinois, can be found below.

Assuming the planet were a globe, let's determine Chicago's location in respect to Grand Mere Park. The elevation of Grand Mere Park is 600 feet. 577 feet above sea level is where Lake Michigan is. Grand Mere Park is thus 23 feet over Lake Michigan since 600 x 577 = 23. Assume for a moment that the photographer was standing in Grand Mere Park's highest location. For the height of the photographer, we will add six feet. Using the formula 23 + 6 = 29, we determine that the camera was, at most, 29 feet above Lake Michigan's surface. To account for the camera's elevation above Lake Michigan, we shall deduct seven miles from the distance of 57 miles (a 29-foot drop to the horizon equals about 6.6 miles, which is rounded up to 7 miles) (29 feet = 6.6 x 6.6 x 8 inches).

Inferring the earth's curvature from the 50-mile radius, we discover that the ground level in Chicago should be 1,644 feet below the horizon. The Sears Tower is the highest structure in the image (it has been recently renamed the Willis Tower). From 1974 to 1998, the Sears Tower held the record for highest skyscraper in the world. It is 1,450 feet above the ground level. The antennae on the tower's summit raise the overall height to 1,729 feet above street level, though.

It must be recognized that in order to calculate the 1,644-foot drop below the horizon, it was required to deduct 23 feet from the 1,667-foot total drop (50 x 50 x 8 inches = 1,667 feet) in order to account for the fact that the Sears Tower is 23 feet above the level of Lake Michigan. 595 feet above sea level, the Sears Tower. There are 577 feet above sea level in Lake Michigan. The Sears Tower is 23 feet over Lake Michigan at that altitude. Since 1,667 x 23 = 1644, the Sears Tower's base would be 1,644 feet below the horizon. Therefore, none of the structures, including the Sears Tower, would be visible if the planet were a globe. Each one of them would be below the horizon. (1,644 divided by 1,450 equals 194), the peak of the Sears Tower would be 194 feet below the horizon. The top 85 feet of the Sears Tower's antennas would be the sole feature of the whole Chicago skyline that could be seen.

In actuality, the Nowicki image clearly shows the whole Sears Tower as well as every other structure along the Chicago shoreline. Joshua Nowicki's image of the Chicago skyline serves as evidence that the planet is flat.

Figure: Sears Tower, showing the tower's height and its position in respect to the horizon if the world were a globe when viewed from 57 miles away across Lake Michigan.

The media had to explain how Chicago could be seen from the Michigan coast, 57 miles distant, which is impossible if the planet were a globe, after Joshua Nowicki's shot caused such a stir. According to ABC News weatherman Tom Coomes, it is not feasible to see Chicago from 57 miles away. Chicago is beyond the horizon; you shouldn't be able to see it, according to Coomes. However, it is visible and evidently so in the image by Joshua Nowicki. What seems to be the apparent impossible does Coomes explain? Coomes said that the Chicago skyline was a fabrication, explaining away Nowicki's image. He asserted that Nowicki's image showed what is referred to as a "better mirage." Coomes displayed incredible restraint by maintaining a straight expression throughout his absurd justification.

Figure: Tom Coomes explaining that Joshua Nowicki's photo of the Chicago skyline was not really there at all; it was a mirage.

A superior mirage is a mirage of an item that is typically reversed over the actual thing, which presents an issue with Coomes' interpretation. A superior mirage happens when an image of an item appears above the real thing, according to Sjaak Slanina. 49 The illustration below depicts the inversion of a superior mirage and is courtesy of the National Snow and Ice Data Center (NSIDC).

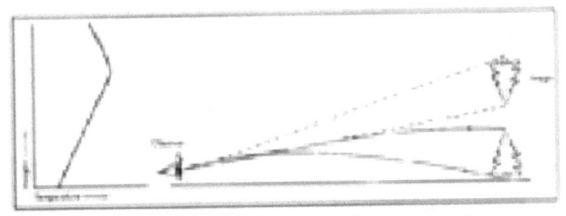

Figure 18: NSIDC Explanatory Diagram of a Superior Mirage

Figure: NSIDC Explanatory Diagram of a Superior Mirage

An genuine superior mirage that appeared over a ship is seen in the image below. The illusion is upside-down and appears above the real vessel, as you can see. Its mirage and the cargo are both visible. The degree of detail in this particular mirage is extraordinary, yet despite this, the mirage is still severely distorted; it fades toward the stern of the ship without showing the back cabin at all.

Is it feasible to view the skyline of Chicago in a better mirage? Yes, it is the solution. The image of a superior Chicago mirage below was captured by J. Michael Hall on April 18, 2015, when he was at St. Joseph on the western edge of Lake Michigan. The Chicago towers are highly warped and blurry, as you can see. The towers' orientation is perhaps most notable. That contrasts sharply with what is seen in Nowicki's image from Grand Mere Park.

Figure: Superior mirage of Chicago skyline taken from St. Joseph, Michigan.

The Chicago skyline in Joshua Nowicki's photograph doesn't resemble a better-quality illusion at all. The skyline is captured in fine detail and upside down in Nowicki's image. The Chicago skyline would most likely be upside-down and noticeably deformed if Nowicki's image were an improved illusion. However, unlike a superior mirage, which causes light to reflect upward, the most typical refraction of light occurs downward. Light is refracted downward by the moisture in the air. In turn, this makes an object's perceived height (what is seen) look less than it actually is. In other words, if the earth were a globe, the likelihood of an item being seen over the curve owing to refraction would be less, not higher. Another impact of the atmosphere's lensing is the magnification of distant objects, which obscures the bottoms of such objects from view. Many people mistakenly believe that the earth's curvature is what accounts for the missing bottoms of buildings and ships; however, they are really created by the lensing effect of the atmosphere. The fact that the seen item frequently cannot be seen at all if the earth were a globe, as it should be entirely below the apparently bent horizon, is evidence that the missing bottoms are caused by the atmospheric lensing effect.

There is little doubt that the skyline of Chicago captured by Nowicki is real. Rick Hummer and Rob Skiba both affirmed that. The Chicago skyline was visible to Rob Skiba and Rick Hummer on June 24, 2016, from the mouth of the New Buffalo Harbor on the Michigan side, a distance of around 42 statute miles. They had leased a boat for the trip. More critically, they practically crossed the full breadth of Lake Michigan from the Michigan side at New Buffalo Harbor to within about 9 statute miles (about 8 nautical miles) of Chicago, proving that the city's skyline was not an illusion.

The image of the Isle of Man below was captured on August 10, 2012, near Rossall Beach in England, looking across the Irish Sea. On the Isle of Man's east coast, Rossall Beach and Port Soderick are separated by around 63 miles. Rossall Beach is 20 feet above sea level. The height of the photographer over the water is 26 feet after adding an additional 6 feet for their height. Assuming a curved earth, the horizon at 26 feet above the ocean won't start to disappear from the photographer's line of sight until they are around 6 miles away. The 63-mile distance between the Isle of Man and the mainland is reduced by 6 miles to 57 miles. The curvature of the earth's 57-mile radius between the photographer and the Isle of Man's beach must be taken into consideration. Rossall Beach is 2,166 feet below the horizon of the Isle of Man, according to the formula miles squared x 8 inches Equals distance below horizon (57 x 57 x 8 inches = 2,166 feet). The highest mountain on the Isle of Man, which rises to a maximum elevation of 2,037 feet, would be 129 feet below the horizon if the planet were a globe.

Figure: Isle of Man seen from Rossall Beach, 63 miles away

One would be curious about the distance between Rossall Beach and the Isle of Man's highest peak given that it is the island's highest point—not its coastline—that is in question. At 2,307 feet above the Irish Sea, Snaefell is the highest peak in the Isle of Man. Also around 63 miles away from Snaefell over the Irish Sea.

About 9 miles offshore from Rossall Beach are the windmills shown in the shot. The windmills appear larger than they actually are in respect to the Isle of Man because they are significantly closer to the camera than the island. The top of the rotors of the windmills may rise 425 feet above the water when they are cranked to the 12 o'clock position.

Figure: Map showing distance from Rossall Beach to Port Soderick on the east coast of the Isle of Man.

Rossall Beach could not possibly have a view of the Isle of Man unless the planet were flat. The planet is flat, as shown in the image above of the Isle of Man taken from Rossall Beach. The picture of the stunning sunset was shot from Rossall Beach and shared on the website of the Rossall Beach Residents and Community Group. The community and residents of Rossall Beach were unaware that the picture depicts more than just a lovely sunset; it is proof that the planet is flat.

The image that follows is a frame from a movie. On Edmonds' Fisherman's Pier, a video camera with a telephoto lens was positioned about 15 feet, 11 inches above the water. Looking north-northeast over Puget Sound, the camera is focused on a ferry that travels between Mukilteo, Washington, and

Clinton, Washington. The boat may be seen in the image approaching Clinton, Washington. Directly across the sea, the distance from Fisherman's Pier to Mukilteo is 10.34 miles. The camera drifts farther away from the boat as it moves from Mukilteo to Clinton. The distance from the camera at Fisherman's Pier to the lengthens to 10.81 miles midway between Mukilteo and Clinton. Approximately 11.3 miles separate the

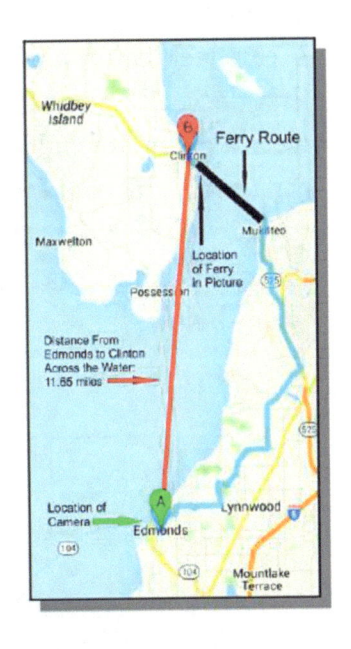

Fisherman's Pier from the point just before arrival at Clinton. According to www.FreeMapTools.com, it is 11.65 miles from Edmonds to Clinton itself in a direct line. We'll round down the distance to 11 miles for simplicity's sake as the ship in the image is about to arrive in Clinton. The full boat shouldn't be seen from Fisherman's Pier if the earth were round, as is popularly thought. The ferry would be 25 feet below the horizon at a distance of 11 miles, after accounting for the camera's height off the dock (we'll round up to 16 feet for simplicity). That is, there shouldn't be any visibility at all in the bottom 25 feet of the ship. But in the image, the white deck cabins contrast sharply with the black topsides of the hull. A car-deck cabin with a 16-foot clearance makes up the first deck. Although it is visible in the image, if the world were a globe, it would be below the horizon. However, we can see the full boat from head to stern, including the black topside of the hull at the water's edge. Simply said, unless the world is flat, it is impossible to view the full ferry in the image below.

Figure: Frame from video of ferry taken from 11 miles away.

Here is a schematic of the Toitea, the actual ferry shown in the image above. If the world were a globe, the horizon should have obstructed around 25 feet. As you can see in the image above, the first deck and the whole black topside of the hull would be covered by 25 feet. The black topside of the ship is eight feet above the sea, according to the ship's builder. About three feet broad is the green stripe. If the world were a globe, the 11 feet that make up the green stripe and the black topsides should be entirely below the curve of the planet. The black topside of the boat, however, is easily seen against the lake.

25

The image below was captured from the Nevi Promenade. December 4, 2007, was the day the photo was shot. In the background of the image lies the Island of Corsica. Monte Cinto, Corsica's highest peak, reaches 8,878 feet above sea level. Nevi is located 139 km from Monte Cinto. Nevi rises 82 feet above sea level at its highest point. We deduct 12 miles from the 139 mile trip to account for the photographer's elevation in Nevi (82 feet above sea level plus 6 feet in height = 88 feet). The distance between the photographer at Nevi and Monte Cinto, the highest peak on Corsica, would be 127 miles if the planet were a globe. The alleged curve would result in a 10,753-foot drop across the 127-mile trip. As a result, the peak of Monte Cinto should be 1,875 feet below the horizon (10,753 8,878 = 1,875) if the world were a globe. But that is not the case. The snapshot shows the whole Island of Corsica, including the summit of Monte Cinto, proving that the world is not a globe. The view of Corsica as seen from Nevi is only possible on a flat earth.

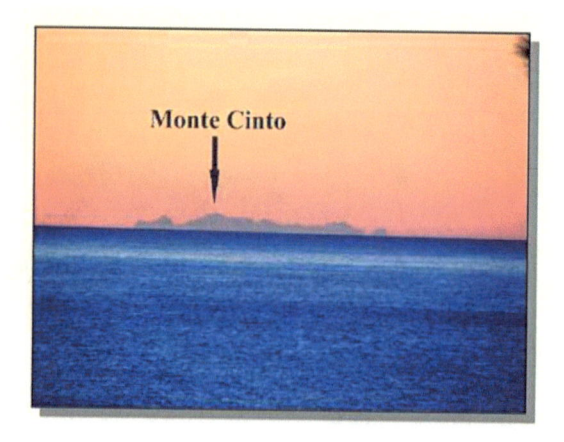

Figure: Photograph of Corsica taken from Nevi, 139 kilometers distant from Monte Cinto, the island's highest point. The top of Monte Cinto should be 1,875 feet below the horizon if the planet were a globe.

A chart of nautical lights and buoys has been made available by the United States Coast Guard. 66 The graph supports the flat earth theory. For instance, according to the U.S. Coast Guard, the Cape Canaveral Lighthouse's focal plane is 137 feet above mean high water. The light's stated maximum range is 24 nautical miles (approximately 28 statute miles). In clear weather, which is defined as having a visibility of at least 10 nautical miles, a light can be seen at a distance called the nominal range.

The light on the lighthouse would be 247 feet below the horizon at a distance of 28 statute miles from the observer, assuming the world is a globe and the sailor is 10 feet above the surface of the water when seeing the Cape Canaveral Lighthouse. We take the 28 statute miles and deduct 4 statute miles to account for the observer's height (10 feet) above sea level (which gives us 24 miles). The result of multiplying the square of 24 miles (576) by 8 inches is 4,608 inches (384 feet) of descent below the horizon on a hypothetically curved globe. The light on the Cape Canaveral Lighthouse would be 247 feet below the horizon, assuming that the observer was 10 feet above sea level at a distance of 28 miles from the lighthouse, when we deduct the height of the lighthouse's focal plane (137 feet) from the horizon drop (384 feet). But it's not 247 feet below the horizon where the lighthouse is. This can only suggest that the earth is not a sphere. The fact that seafarers can see the Cape Canaveral Lighthouse from 28 statute miles away is evidence that the planet is flat.

A Geographic Range Table that provides "the approximate geographic range of visibility for an item which may be observed by an observer at sea level" is included in the publication from the United States Coast Guard previously mentioned. The table is displayed below. If the world were a globe, the objects at the stated height would not be seen from the distances listed in the table. For instance, the table shows that an observer at sea level may see an item that is 200 feet above the surface from a distance of 16.5 nautical miles (19 statute miles). The top of that 200-foot item would be 40 feet below the horizon from 19 statute miles distant if the planet were a globe (192 x 8 inches = 240 feet). The range table is accurate since it is based on testing and experience from the actual world. Only if the earth is flat could the objects listed in the range table be seen from the distances on the chart. The range table provides verified proof that the earth is flat from the United States Coast Guard.

GEOGRAPHIC RANGE TABLE

The following table gives the approximate geographic range of visibility for an object which may be seen by an observer at sea level. It is necessary to add to the distance for the height of any object the distance corresponding to the height of the observer's eye above sea level.

Height Feet / Meters	Distance Nautical Miles (NM)	Height Feet / Meters	Distance Nautical Miles (NM)	Height Feet / Meters	Distance Nautical Miles (NM)
5/1.5	2.6	70/21.3	9.8	250/76.2	18.5
10/3.1	3.7	75/22.9	10.1	300/91.4	20.3
15/4.6	4.5	80/24.4	10.5	350/106.7	21.9
20/6.1	5.2	85/25.9	10.8	400/121.9	23.4
25/7.6	5.9	90/27.4	11.1	450/137.2	24.8
30/9.1	6.4	95/29.0	11.4	500/152.4	26.2
35/10.7	6.9	100/30.5	11.7	550/167.6	27.4
40/12.2	7.4	110/33.5	12.3	600/182.9	28.7
45/13.7	7.8	120/36.6	12.8	650/198.1	29.8
50/15.2	8.3	130/39.6	13.3	700/213.4	31.0
55/16.8	8.7	140/42.7	13.8	800/243.8	33.1
60/18.3	9.1	150/45.7	14.3	900/274.3	35.1
65/19.8	9.4	200/61.0	16.5	1000/304.8	37.0

Example: Determine the geographic visibility of an object, with a height above water of 65 feet, for an observer with a height of eye of 35 feet.

Enter above table:
Height of object 65 feet= 9.4 NM
Height of observer 35 feet= 6.9 NM
Computed geographic visibility= 16.3 NM

Figure: United States Coast Guard Geographic Range Table

Made in the USA
Columbia, SC
09 July 2025